Future of Business and Finance

The Future of Business and Finance book series features professional works aimed at defining, describing and charting the future trends in these fields. The focus is mainly on strategic directions, technological advances, challenges and solutions which may affect the way we do business tomorrow, including the future of sustainability and governance practices. Mainly written by practitioners, consultants and academic thinkers, the books are intended to spark and inform further discussions and developments.

More information about this series at https://link.springer.com/bookseries/16360

Carlo Bagnoli • Andrea Albarelli •
Stefano Biazzo • Gianluca Biotto •
Giuseppe Roberto Marseglia •
Maurizio Massaro • Matilde Messina •
Antonella Muraro • Luca Troiano

Digital Business Models for Industry 4.0

How Innovation and Technology Shape the Future of Companies

Springer

Carlo Bagnoli
Department of Management
Ca' Foscari University of Venice
Venice, Italy

Andrea Albarelli
Department of Environmental Sciences, Informatics and Statistics
Ca' Foscari University of Venice
Mestre, Venice, Italy

Stefano Biazzo
Department of Management and Engineering
University of Padua
Padua, Italy

Gianluca Biotto
Strategy Innovation S.r.l.
Venice, Italy

Giuseppe Roberto Marseglia
University of Pavia
Pavia, Italy

Maurizio Massaro
Department of Management, Ca' Foscari University of Venice
Venice, Italy

Matilde Messina
Strategy Innovation S.r.l.
Venice, Italy

Antonella Muraro
Avanade Italy S.r.l.
Milan, Italy

Luca Troiano
Zeb Consulting S.r.l.
Milan, Italy

ISSN 2662-2467 ISSN 2662-2475 (electronic)
Future of Business and Finance
ISBN 978-3-030-97283-7 ISBN 978-3-030-97284-4 (eBook)
https://doi.org/10.1007/978-3-030-97284-4

This Springer imprint is published by the registered company Springer Nature Switzerland AG
The registered company address is: Gewerbestrasse 11, 6330 Cham, Switzerland

Foreword

Francesco Caio is the chief executive officer of Poste Italiane S.p.A. and of Saipem S.p.A., an Italian multinational oil organization operating in the service sector, leading company in the provision of engineering services and in the realization of important projects in both the energy and infrastructure industries.

Digitalization is transforming business activity and structure in all sectors: industry, services and finance.

The breadth and depth of the impact are not easy to understand and describe; often many textbooks and speeches on the subject consider specific companies, market segments or types of transactions.

This makes it complex for students and economic actors not only to grasp all the aspects and implications of the current digital revolution but also to acquire a taxonomy of the phenomenon, which is essential for analysing and, above all, governing the process that, more than any other, is affecting and will affect the competitiveness of every company.

In this context, the textbook represents a very useful atlas to define the boundaries of this vast ocean and plough through the ocean currents with full knowledge of the facts.

The concept of business model offers an effective cognitive platform for at least two orders of reasons. On the one hand, it renders well the idea that digitalization is not an "aesthetic" manoeuver of modernization of the enterprise—for example, in its relationship with clients—but a structural (re)design of its foundations. Digitalizing means rethinking, disassembling and rebuilding processes, capital allocation, fixed assets and working capital. On the other hand, it helps to understand that—since no company is a monad—the digital business models modify the relationships between companies and between supply chains with deep implications on the processes of "engagement" with suppliers and customers but also, and in some cases above all, with distribution systems and "route to market".

Digital technologies can radically transform processes and products. As a result, companies need to embrace the opportunities offered by the digital revolution to strategically innovate their business, generate new revenue streams and remain competitive in a constantly changing market.

The power of data and information in the era of the Fourth Industrial Revolution has contributed to the development of new business models, proposing disruptive alternatives aimed at innovating the way companies produce and distribute their products and services.

The present work focuses on the identification of four distinct digital business models: data-driven, platform, smart factory and servitization. The adoption of a data-driven business model allows enterprises to gain insights through the creation of user-product interactions to improve product performance and offer a complete customer experience. Multisided platform business models aim to facilitate interactions and transactions between two or more interdependent groups, that is, consumers and producers, by stimulating the creation of network effects, for which the platform gains value as the number of participants increases. Smart factories are characterized by the automation of production processes to promote the large-scale production of personalized goods. Service-centric companies activate a transformative process, whereby products previously manufactured and sold are converted into services to strengthen relationships with consumers, ensuring recurring revenue streams.

The work finds an effective balance between the definition of a reference model and its application to real-world business cases that have already been able to realize it with important reflections on their competitiveness, growth and profitability.

The book is therefore not only a study guide or an in-depth study but a viaticum for those—entrepreneurs and managers—who have grasped the centrality of digitalization and have started, or have decided to start, the transformation necessary to translate its potential into results and value.

Obviously, like all transformations, digitalization requires an act of courage and vision, as well as the determination and discipline to carry it through successfully.

Now, as Manzoni reminds us with Don Abbondio, courage is a propensity of the spirit that can only come from a deep internal conviction; but for those who have dared to digitalize, the textbook offers an excellent guide to do it well, with strategic and organizational awareness.

Naples, Italy Francesco Caio

Introduction

The Fourth Industrial Revolution, also referred to as Industry 4.0, primarily and significantly reduces the positive effects of economies of scale. Small-sized companies shift from representing a disadvantage to becoming a source of advantage involving no longer the size of the individual organization and yet instead its ability to position itself within the most suitable strategic area, according to its value (eco)system. Digital transformation alters the modalities through which business is conducted following the introduction of advanced solutions allowing companies to reimagine their role, therefore impacting their entire value (eco)system, from product design, including the products' embedded intelligence which allows to manage their entire life cycle, to both supply and sub-supply relationships, allowing real-time data processing, and from production processes managed as cyber-physical spaces to logistics as well as storage systems and to direct physical and digital contact with end customers, as the boundaries between the supply of goods and services are becoming increasingly blurred.

The ongoing industrial revolution therefore opens opportunities for companies in terms of process efficiency, cost reduction and productivity enhancement by enabling large-scale product customization. Nevertheless, it also allows to rethink products, to introduce new pre- as well as post-sales services and to improve the ability to respond to market needs. The technology push innovation, characterizing Industry 4.0, enables companies to exploit their potential in order to organize, integrate and regulate their corresponding production chains. This implies a deviation from a fragmented model in favour of an interconnected supply chain model in which organizations collaborate for the purpose of providing a competitive output. The cooperation among companies in the form of networks is also going to grant organizations' access to financial, technological resources and, more generally, to sources of knowledge.

However, the Fourth Industrial Revolution will contribute to the radical competitive repositioning of companies under a specific condition: the opportunities offered by the multiple digital paradigms characterizing Industry 4.0 must be exploited for the purpose of designing innovative digital as well as functional business models. Moreover, the technological innovation characterizing production processes and products must be accompanied by strategic business model innovation. This may reduce costs through the elimination of critical success factors but at the same time increase revenues due to the introduction of new competitive factors. As a result, the digital revolution might contribute to creating new markets starting from the fulfilment of both emerging customer needs and social needs, as the consolidated profit models are unable to satisfy them. Therefore, Industry 4.0 enables the redefinition of business models within all sectors.

This research aims to investigate the potential impact of Industry 4.0 on business models, considering technological innovation as an additional driver of strategic innovation. To this end, a structured literature review was conducted, examining a total of 182 documents (selected among scientific papers, consultancy and institutional reports), devoted to the topic of Industry 4.0 and published between 2010 and 2019. This research has made it possible to quantify the importance of the topic, considering the potential impacts of the following:

1. The digital paradigms characterizing Industry 4.0 on the individual building blocks constituting the numerous business models illustrated within the research, thus leading to the identification of multiple operational benefits
2. The operational benefits identified based on the critical success factors underlying the value propositions and on the strategic dimensions, leading to the identification of 22 business models grouped into 4 macro categories

The first chapter explores the issue of strategic innovation as well as the sources of innovation. Industry 4.0 represents a significant driver for the development of new business models, enabling companies to change the rules to their advantage. The first chapter also describes the framework designed in order to identify the potential impacts of the digital paradigms characterizing Industry 4.0 on the numerous building blocks and therefore on the business models.

The second chapter illustrates the nine technological paradigms characterizing Industry 4.0 (including additive manufacturing, autonomous and collaborative robots, augmented and virtual reality, cloud computing, simulation, Industrial Internet of Things (often shortened as IIoT), big data analytics, cybersecurity and structural horizontal and vertical system integration) for the purpose of describing their characteristics as well as of highlighting their potential impacts on the building blocks constituting the business model canvas. Nine tables are presented, focusing on the operational benefits generated by each digital paradigm in accordance with the literature.

The third chapter identifies 22 innovative business models by combining the potential impacts of the operational benefits identified based on the critical success factors underlying the value propositions and on the strategic dimensions: indirect data monetization, product endless improvement; indirect data monetization, customization; indirect data monetization, product upgrading/up-selling; indirect data monetization, new product development/cross-selling; indirect data monetization, new market space creation; direct data monetization; data confederation; infrastructure community; social network; marketplace; smart factory; decentralized production and control; smart supply chain; agile collaboration network; hub and spoke; production as a service; product ownership; product + service (add-on hardware);

product + service (add-on software); product availability; product as a service; and product subscription. The aforementioned business models are grouped into four meta business models, so as to identify the following macro categories: data-driven, platform, smart factory and servitization.

In summary, the research illustrates that if the opportunities offered by the Fourth Industrial Revolution are fully exploited, organizations are going to be able to combine the characteristics of the digital transformation within their business structure and will no longer chase the competition; instead, they are going to contribute to changing the rules of the game. Therefore, individual companies as well as their corresponding production chains will have to focus on process as well as product innovation following the advent of Industry 4.0 and especially on strategic business model innovation. The strategic challenge is not to improve the same matter but to do different things or do things differently.

[illegible]

Contents

1 Strategic Innovation Driven by Digital Transformation

1.1 Strategic Innovation

Since 2020, companies have suffered severe financial difficulties following the latest COVID-19 health and economic crises, which accelerated the digital evolution that was already underway in the competitive context [1]. Even before the arrival of the crises, companies faced critical strategic challenges as a result of emerging:

- *political trends*, primarily the introduction of protective tariffs imposed by the United States and designed to shield domestic production from foreign competition, especially considering the U.S.–China strategic competition. The former represents the answer of the Trump administration to China's Belt and Road Initiative;
- *economic trends*, mainly the globalization of production and markets which increases the intensity of competition and reduces the product life cycle, thus gradually transforming products into commodities and eroding profits;
- *socio-cultural trends*, mostly the changes in consumer behaviour, as consumers are becoming increasingly conscious of the importance of certain aspects of production, including creativity, design, as well as sustainability, therefore seeking consumption experiences personalized to them;
- *technology trends*, primarily the advent of the Fourth Industrial Revolution, contributing to the introduction of automated production processes and to the reconfiguration of smart products, production processes, and value chains, therefore significantly reducing the positive effects of economies of scale;
- *environmental trends*, in particular increased awareness concerning unsustainable production, as well as consumption models, thus contributing to the identification of the core principles of the circular economy; and
- *legal trends*, primarily the General Data Protection Regulation (GDPR), that is the European regulation on data protection which applies to all organizations storing personal information about citizens and companies within the European

C. Bagnoli et al., *Digital Business Models for Industry 4.0*, Future of Business and Finance, https://doi.org/10.1007/978-3-030-97284-4_1

Union and further countries. In addition, the regulation seems to encourage organizations to develop data-driven business models [1, 2].

The global financial and economic crisis has accentuated the necessity for most Italian companies to increase investments in innovation, regardless of whether to pursue a particular strategy aimed at enhancing:

- *Productivity*, for instance reducing unit production costs, and therefore improving the cost structure or reducing direct as well as indirect costs, and/or increasing the utilization of assets, thus reducing fixed and working capital;
- *Growth*, for example increasing total sales revenue from existing customers or increasing the sale of products that have already been purchased by customers or that have yet to be purchased and/or increasing revenue streams through the creation of new products or market space. [3].

Innovation, defined by Joseph Alois Schumpeter as the practical application of an invention, represents the major driver for creating a competitive advantage. Innovations can be distinguished according to:

- The *object of innovation*: product innovations (introducing better or new goods and services) vs. process innovations (introducing improved or new ways to produce or distribute goods and services);
- The *intensity of the effects*: incremental innovations (improving existing goods and services) vs. radical innovations (creating new goods and services);
- The *effect on performance*: supportive innovations (improving 'traditional' performances, meaning those appreciated by the majority of customers (e.g. power) vs. disruptive innovations (as 'traditional' performances worsen, organizations must improve those initially appreciated only by a minority of customers—usually extreme users—or by noncustomers (e.g. small).
- The *effect on skills*: competence enhancing innovations (re-evaluating existing skills) vs. competence destroying innovations (devaluing existing skills);
- The *development methodologies*: modular innovations (modifying the components of the system) vs. architectural innovations (modifying the configuration of the relationships amongst the components of the system).

Nonetheless, the distinctions aforementioned tend to intersect with one another. In particular, incremental innovations are typically competence enhancing, whilst radical innovations are generally competence destroying. Henderson and Clark proposed a framework to define the different types of product innovations which contribute to integrating the distinctions based on the intensity of their effects, as well as on skills, with the development methodologies as illustrated in Fig. 1.1 [4].

The objects of the in-depth innovation illustrated in this research are the organizational strategy and the business model. Renovating the organizational strategy requires to overcome the competitive market paradox, therefore increasing product value and lowering production costs, thus introducing a new value proposition and

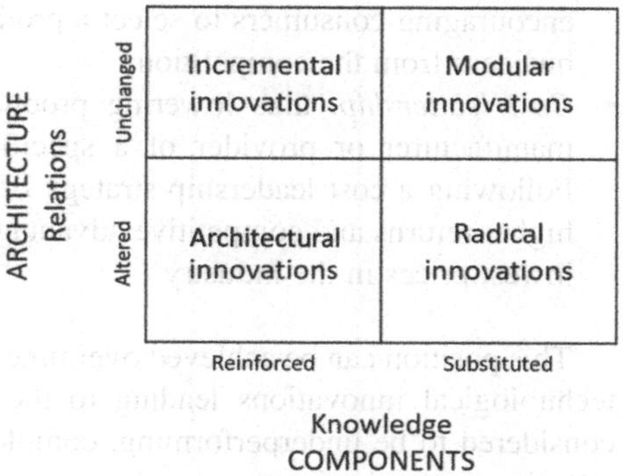

Fig. 1.1 The framework to define product innovations

creating a new market space. In addition, renovating the organizational strategy implies a proper business model redefinition. Furthermore, the global financial and economic crisis has accentuated the necessity for most companies to redefine their business models. Some companies may limit themselves to perfecting their existing business models, however most organizations, especially those (mainly Italian enterprises) offering artisan custom creations and characterized by strategic pricing, both constituting important competitive elements, are forced to change their business models so as to implement a strategic innovation [4].

However, what is a strategic innovation? And, what are the steps involved in the development of a strategic innovation?

1.1.1 Strategic Innovation: Features and Processes

There are multiple definitions of strategic innovation which seem to converge on the fact that strategic innovation results in the design of a new business model [5]. The latter can be achieved through the development of:

- Innovative products: (goods and/or services), presented or combined in a new way, so as to create a radically different and transformative experience for customers, emotionally, intellectually, and/or spiritually engaging customers;
- Innovative processes: production processes and/or distribution of existing or new products which contribute to acquiring new customers;
- Innovative value chains, so as to create a new market space which, assuming competition is irrelevant, allows for an increase in value, thus benefiting both the company and its customers.

The first challenge consists of overcoming the pursuit of competitive strategies, including:

- *Differentiation*, therefore increasing perceived value as well as raising prices, leveraging on the business's competitive advantage over competitors and encouraging consumers to select a product over another, thus differentiating the business from the competition;

- *Cost leadership*, thus lowering production costs and becoming the cheapest manufacturer or provider of a specific product or service in a competition. Following a cost leadership strategy offers organizations the possibility to earn higher returns and competitive advantages by offering products or services at the lowest prices in the industry.

This position can be achieved over time through the implementation of disruptive technological innovations leading to the development of new products that are considered to be underperforming, complex, bulky and, therefore, expensive, thus allowing organizations to achieve cost leadership. Subsequent supportive innovations will lead to increased product performance. Furthermore, innovation surpasses the rate at which customer needs evolve. Therefore, the products will be characterized by traditional performances such that the majority of customers will no longer be able to appreciate or will not be willing to pay a surcharge for them. The new products, on the other hand, will be characterized by traditional performances capable of satisfying the majority of customer needs, offering value to customers, thus achieving a differentiation-based competitive advantage [6]. In order to achieve an organizational competitive advantage, products must become industry standards, thus generating high switching costs for customers. The present lock-in strategy contributes to building a strong network of customers and to favouring the creation of complementary products, therefore discouraging customers from switching to competitors [7].

Both of the abovementioned strategies are based on 'first-mover' benefits [8].

In order to overcome the strategic paradox between value and cost, Kim and Mauborgne propose to utilize the strategic framework, constituting of a graphical representation in which the attributes of value or the critical success factors in a given market are displayed on the horizontal axis, while the vertical axis depicts the level of performance achieved by the business and its direct competitors (see Fig. 1.2) [5]. The value curve is obtained by connecting the performance levels achieved by the company considering its multiple critical success factors. Comparing the value curve of the organization with that of its direct competitors allows the company to understand which value attributes are higher and which are lower.

The pursuit of a differentiation-based competitive advantage would lead the company to focus on its weaknesses in order to reduce the bad effects of competition, therefore offering similar or even slightly better value than competitors. However, this strategy would lead to a significant increase in production costs, risking that the company's products would be priced out of the market. Pursuing a cost leadership strategy, on the other hand, would lead the organization to divest of its strengths in order to reduce the good effects of competition, thus offering similar or even slightly better value than competitors. The present strategy aims at reducing production costs and winning price wars. Finally, pursuing an innovative strategy would lead the enterprise to widen the competitive gap, shifting the focus from current customers, properly segmented in order to identify the target groups, as well

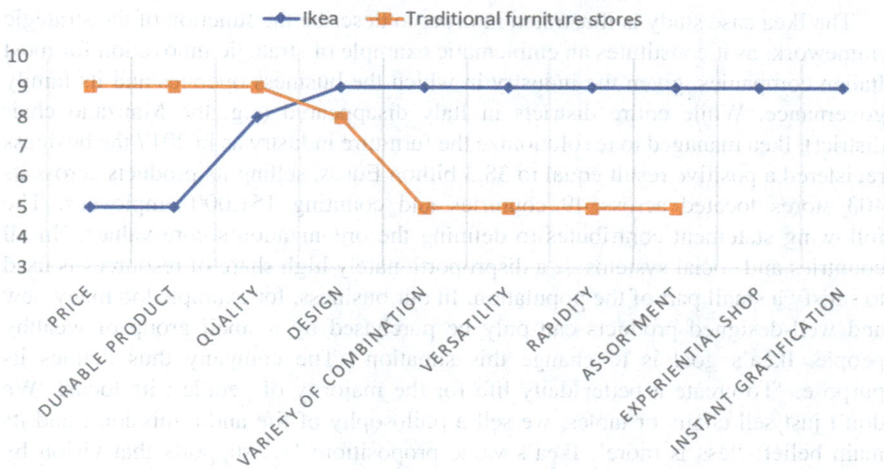

Fig. 1.2 Ikea's value proposition: the value curve

as to increase the market share, to non-customers adequately connected in order to identify similarities between customers for the purpose of creating a new market. As a matter of fact, an innovative strategy aims at providing disruptive answers to the following questions: 'What critical success factors…

- …could be increased above industry standards for the purpose of increasing product value, as customers consider them to be key value attributes?
- …although not required according to industry standards could be created for the purpose of increasing product value, as customers consider them to be key value attributes?
- …required according to industry standards could be eliminated, therefore reducing production costs? Moreover, why are customer attributes not taken into consideration?
- …could be reduced below industry standards, for the purpose of reducing production costs? Moreover, why are customer attributes not considered important?'

Successful strategic innovation should lead to the design of a value proposition focused on a few critical success factors rather than sustaining excessive production costs or failing to engage customers, as well as noncustomers, and therefore distinguishing the business from the competition. The core attributes which the company has chosen to invest in or not to invest in should be logically grouped for the purpose of avoiding obtaining a 'low-high-low-high-low-high' pattern characterizing the value curve. Indeed, this representation would betray the lack of a coherent strategy, as the result of a mere juxtaposition of various partial strategies that fail to follow a particular strategic vision.

The Ikea case study is illustrated in order to describe the function of the strategic framework, as it constitutes an emblematic example of strategic innovation for most Italian companies, given the industry in which the business operates and its family governance. While entire districts in Italy disappeared (e.g. the Manzano chair district), Ikea managed to revolutionize the furniture industry as in 2017 the business registered a positive result equal to 38.3 billion Euros, selling its products across its 403 stores located across 49 countries and counting 151,000 employees. The following statement contributes to defining the organization's core values: 'In all countries and social systems… a disproportionately high share of resources is used to satisfy a small part of the population. In our business, for example too many new and well-designed products can only be purchased by a small group of wealthy people. Ikea's goal is to change this situation'. The company thus defines its purpose: 'To create a better daily life for the majority of people'; its focus: 'We don't just sell chairs or tables, we sell a philosophy of life and a mission'; and its main belief: 'less is more'. Ikea's value proposition: '… supports that vision by offering a wide range of functional and design home furnishings at prices so low that as many people as possible can afford them'.

Figure 1.2 compares the Ikea value curve with that of traditional furniture stores. The comparison of the two value curves graphically illustrates how Ikea's value proposition differs from that of traditional furniture stores, ignoring the attributes once considered as significant (e.g. durable product) and vice versa (e.g. variety combinations), as well as introducing new attributes such as experiential shop and instant gratification. The mirror symmetry between the two curves highlights the strategic innovation operated by Ikea.

In general, the first step in order to start a strategic innovation consists in providing an answer to the following questions for the purpose of (re) defining the business (model):

1. Who are our customers?
2. What do we offer them?
3. How can we deliver it effectively and efficiently?

These questions are connected to three strategic dimensions (Fig. 1.3):

1. *Customer intimacy*: pursuing a strategy of customer intimacy requires to quickly and accurately respond to the specific requests of a niche of customers (identified as a result of market segmentation), offering customers' tailored experiences;
2. *Product leadership*: pursuing product leadership requires to fulfil customer needs, offering them innovative products characterized by new and/or higher performance features;
3. *Operational excellence*: pursuing operational excellence requires to fulfil customer needs, offering them standardized products and distributing them, thus removing barriers to usage.

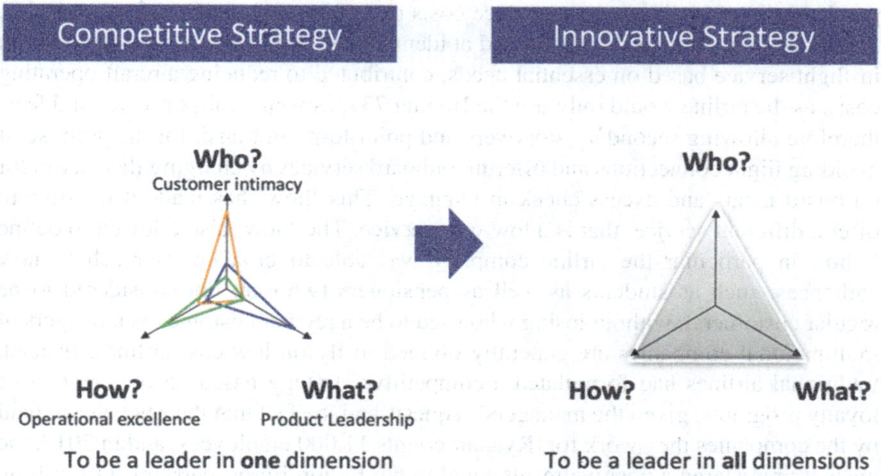

Fig. 1.3 Shifting from competitive strategy to innovative strategy

Competitive strategies are required to define the business in order to formulate the strategy and to answer the above questions in a 'traditional' way. The goal is to become leader in one of the aforementioned strategic dimensions, thus achieving competitive parity in the remaining dimensions. The goal is to identify an attractive industry and to position the business in order to win the competition. Innovative strategies require to redefine the business model, as well as to create a new business or market space in order to formulate the strategy and to answer the above questions in an 'unusual' way. Following this perspective, companies position themselves as leaders in all three strategic dimensions. This leads to the identification of a further strategic dimension referred to as ecosystem leadership. The aforementioned strategy is pursued by the keystone company of a business ecosystem. Once an organization has identified a gap within the strategic positioning map and covers it, its idea evolves eventually into a mass market, and a strategic innovation is ultimately realized. The goal is 'to change the rules of the game' [9].

In order to redefine the business model by providing an 'unusual' answer to the above questions, we tend to define first 'Who', then 'What', and finally 'How'. However, defining the target market and determining customer needs first may not lead to business model innovation. Usually if companies were to ask their customers which products they would want them to produce, customers would most likely ask for enhanced products that are high performance and for favourable purchasing conditions. In this regard, Henry Ford's quote is worth mentioning: 'If I had to give my customers what they asked me I would not have produced cars, but larger carriages'. Strategic innovation usually works in the opposite order, that is, it defines first 'How', then 'What', and finally 'Who'.

Perhaps the best-known example of strategic innovation is Ryanair. The Irish airline has exploited the advent of the Internet in order to allow its customers to purchase tickets online so as to reduce costs by cutting the number of ground staff. Pursuing an innovative strategy aimed at identifying 'how' the business can offer an in-flight service based on essential needs, contributed to reducing aircraft operating costs, as the airline would only use the Boeing 737, as well as airport taxes and fees, therefore allowing secondary stopovers and point-to-point transit for the purpose of avoiding flight connections and offering onboard services by charging fliers extra for on-board meals and excess check-in baggage. This 'how' has made it possible to offer a different service, that is a low-cost service. The 'how' also allowed to define 'who', in particular the airline company was able to broaden its reach to new audiences, such as students as well as pensioners (who are not considered to be regular customers), without losing who used to be a regular customer as managers of multinational companies are generally obliged to fly on low-cost airlines. Instead, traditional airlines had formulated a competitive strategy based on the creation of loyalty programs, given the managers' requests and the fact that their tickets are paid by the companies they work for. Ryanair counts 11,000 employees, and in 2017, the business registered a positive result equal to 6.6 billion Euros, carrying 117 million passengers on 413 aircrafts [10–12].

1.1.2 Strategic Innovation: Sources

In order to formulate a strategic innovation for the purpose of (re)defining the business (model), companies could begin by providing an answer to one of the three questions, thus generating three different sources of strategic innovation. Likewise, identifying 'how' leads generally to a *Technology Push* innovation; identifying 'what' generates a *Design-Driven* innovation; in the same manner, determining 'who' leads to a *Market Pull* innovation [13].

Defining the target market enables companies to pursue a market pull innovation, originating from problem identification or need recognition, namely consumer unfulfilled needs. Typically, organizations design structured questionnaires in order to understand consumer behaviour. The analysis of the results determines the value curve, which establishes whether the company is pursuing a competitive advantage or a competitive disadvantage. Pink Taxi, for example is a taxi service operating in several countries from Mexico to Russia in which both drivers and customers are exclusively women. The aim is to prevent both parties from being subjected to sexual harassment. The service has been highly successful despite its low market share. Solutions like this are usually referred to as incremental innovations because customers are rarely able to express needs that go beyond their usual consumer experience [12]. Customers typically ask for enhanced products that are high performance and available at the lowest prices. However, radical innovations can be achieved by:

- Shifting the focus from products to the reason why customers buy those products, thus identifying:
  - the *job to-be-done*, i.e. what they are trying to achieve;
  - the *gains*, i.e. the advantages they wish to achieve; and
  - the *pains*, i.e. the risks and obstacles they face while carrying out their tasks.
- Shifting from the traditional market research conducted through interviews, to the observation of customer behaviour and in particular examining the behaviour of 'extreme' users, who present greater sensitivity to progress than the others and customer needs that most individuals generally do not have. This is achieved through the ethnographic method, which allows to gain a deeper understanding of the phenomenon in question. The focus is not on what the user says, but on what he/she does, thinks, tries, sees, and hears for the purpose of identifying unexpressed needs. In accordance with the diffusion of innovation theory developed by Everett Rogers, the extreme user concept overlaps with Rogers' innovators (frequency distribution: 2.5%) and early adopters (frequency distribution: 13.5%) [10].

Developing innovative products or services allows companies to pursue a design-driven innovation, therefore providing a unique meaning to existing goods and encouraging product redesign. The final aim is to give a new sense to the product. The Kartell Bookworm, for example was designed to take different shapes, offering customers the opportunity to develop their aesthetic taste, as well as to display their intellectual capacity, placing the books that they wish to show. The meaning behind it is much more similar to that of paintings rather than that of a traditional wooden bookcase, holding plenty of books. Design-driven innovations can be both incremental and radical. They are incremental when the company adopts a new language in order to communicate product value, strengthening or refining the meaning attributed to the product. They are radical when the newly adopted language implies either a substantial change in the current meaning of the product or the creation of new meanings. Hence, the client plays an active role in attributing a meaning to the product, both as an individual (emotionally and psychologically) and as part of society (symbolically and culturally). Users have given an opposite meaning to the Volkswagen Maggiolino compared to that originally imagined by Hitler and interpreted by Porsche, considering that it has become a symbol of the Beat Generation. However, these are proposal innovations that arise from an interpretation of the future evolution of socio-cultural models made by designers, and not as a response to the current needs of customers, whether they are explicit or not.

Optimizing production processes enables enterprises to pursue a technology push innovation, leading to new product development for the purpose of satisfying previously unidentified consumer needs. Dyson's Air multiplier^{TD} technology, for instance, provides a smooth flow of air without using traditional ceiling fans. It represents therefore a 'proposal' innovation in which the main challenge is to educate users and make them appreciate the characteristics of the new product. They are typically radical innovations that present a novelty and achieve differentiation, thus significantly improving their performance and introducing new functions.

According to the traditional technology life cycle, organizations initially compete on product functionalities, then on reliability, on the simplicity of use, on the availability, as well as on the integration, and eventually on its price; as for design-driven innovations, the opposite happens. This process turns technology to push innovations into a complex strategic advantage that organizations can manage as it generates a competitive advantage that is not sustainable in the long-term while often requiring long development times, large investments, and new technological, as well as organizational skills. Finally, we can distinguish between modular and architectural technological innovations. The former consists in changing one or more components of a product without reconfiguring the structure of the system or the way in which the components interact with each other. The latter, on the other hand, consists in reconfiguring the structure of the system without changing its components. However, architectural innovations often impose changes also in terms of product components [4].

As previously stated, a sustainable competitive advantage can be achieved by integrating all the sources characterizing strategic innovation. Accordingly, people buy products not only because of their functional characteristics but also because of product messaging. Likewise, products have a dual nature:

- The technological–utilitarian dimension that has to do with functionality and performance, but also with reliability, simplicity, and price.
- The linguistic–symbolic dimension, which concerns messages and meanings, design, emotions, culture, and status.

If the company is aware of the utilitarian and symbolic nature of its products, it can have a complete view of the product by understanding how it can satisfy practical needs, as well as how it can convey a specific meaning on a cultural level. Any product innovation, as well as process or business model innovation, should therefore have both a technical and a semantic content. Changing the language used to communicate product value to customers can be easier if this change is accompanied also by performance as well as functionality improvements. In addition, a technological innovation can be easily achieved if it is accompanied by a change in the meaning of the product [14].

An emblematic example of strategic innovation occurring by integrating the various sources of innovation is that of the Italian company *Foscarini*, which defines its business model as 'The factory without a factory'. The company designs, assembles, and distributes over 55 lines of decorative lamps, meaning over 200 different models, collaborating with more than 33 designers. *Foscarini* utilizes more than 20 materials and 18 innovative processing technologies, partnering with a local ecosystem of suppliers. This allows the company to select the best fit between the technical solution and the innovation proposed by the designers. The strategic choice of not owning a factory therefore depends on the desire to increase the degree of freedom in creating the lamps, in order to preserve the original message of the designer. This business model therefore makes it possible to maximize the value created for customers, designers, but also for *Foscarini*'s suppliers. The suppliers, coherently with the northeastern entrepreneurial culture, are characterized by having important productive skills but poor marketing skills: They know how to produce everything, but they don't know what to produce. *Foscarini*, also thanks to an international network of retailers that allows them to identify latent consumer needs, can determine what to produce in order to make the most out of their technological skills.

1.2 The Business Model Framework

There are multiple definitions of strategic innovation which seem to converge on the fact that strategic innovation results in the design of a new business model. Similarly, there are many definitions of business model, stating that 'it is a conceptual framework to provide a holistic but abstract understanding of the underlying business logic of an organization'. Starting from the triangular figure previously illustrated (Fig. 1.3), a more complex framework has been elaborated. If the 'what' relates to the offer of products as in the initial framework, the 'how' articulates into the resources and into the organizational processes which are functional to the realization of the products and the 'who' extends to the customers, as well as to the suppliers and the company itself and all corporate stakeholders.

The proposed framework (Fig. 1.4) has a triangular shape in order to identify closed economic cycles, but also the importance of strategic innovation processes in terms of consistency and coherence. It is made up of seven building blocks connected by seven relationships [15].

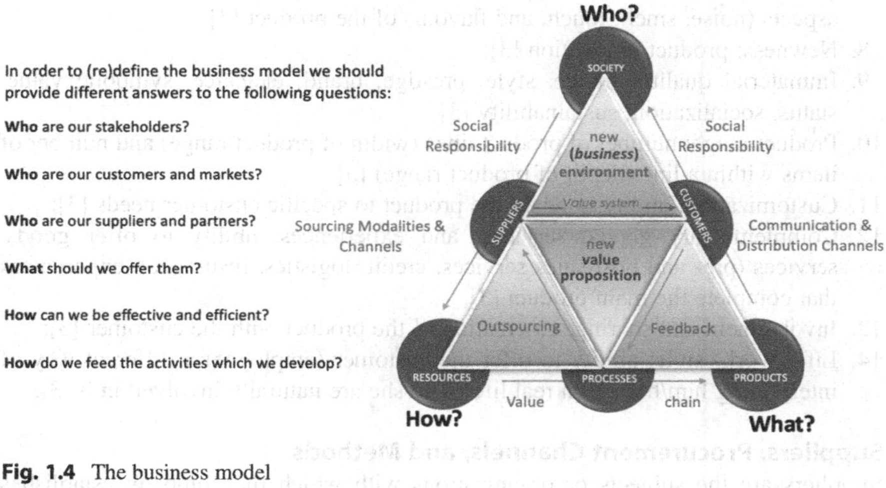

Fig. 1.4 The business model

1.2.1 The Business Model and its Building Blocks

Value Proposition

The value proposition defines how the business produces earnings and/or alleviates customer pains as well as suppliers and further corporate stakeholders' issues, allowing them to accomplish their tasks. An effective value proposition, as already stated, must focus on a few core attributes; other than those offered by competitors or explicitly requested by customers, but perhaps similar to those characterizing alternative industries and requests from non-customers, summarized in a compelling and impactful slogan. The absence of slogans indicates self-referential attitudes and confused value propositions. An effective value proposition must also be deeply rooted in the corporate culture and must necessarily be part of the mission as well as the strategic vision of the company [5, 16–18].

The core attributes, or critical success factors in a given market, can be grouped as follows:

1. Price: purchase cost of the product for the customer [3];
2. Reliability: conformity of the goods produced and respect of delivery times [3];
3. Availability: production capacity to respond to peak demand and warehouse capacity [3];
4. Accessibility: distribution capacity to allow customers to acquire the product anywhere and anytime [3];
5. Rapidity: time needed to deliver the product to the customer once the order is issued [3];
6. Material quality: quality of raw materials and technology, health, safety [3];
7. Performance: functionality, consumption and durability, ergonomics and comfort, simplicity and usability, dimensions and weight, aesthetics and sensorial aspects (noise, smell, touch, and flavour) of the product [3];
8. Newness: product innovation [3];
9. Immaterial quality: image, style, prestige, brand, elegance, symbolic value, status, socialization, sustainability [3];
10. Product range: number of product lines (width of product range) and number of items within a line (depth of product range) [3];
11. Customization: ability to adapt the product to specific customer needs [3];
12. Complementary goods, services, and experiences: ability to offer goods, services (pre- and post-sales services, credit, logistics, insurance), experiences that complete the main product [3];
13. Involvement and sharing: co-creation of the product with the customer [3];
14. Life (work) style: ability to offer the customer (employee) a different way of interpreting him/herself in real life as he/she are naturally involved in it [3].

Suppliers, Procurement Channels, and Methods

Suppliers are the subjects or organizations with which the company establishes relations for the procurement of resources that are necessary to the organizational processes and that are not available internally. Depending on the material *criticality of the resource* to be acquired, that is how much it contributes to the value of the final product and whether or not it is replaceable, and of its *availability on the supply market*, that represents the difficulty with which it is possible to identify alternative suppliers, supply channels can be divided into the following categories:

- *Suppliers on the market*: manufacturers of materials and components that can be distinguished into two categories [19]:
  - occasional: if only the cost of procurement, availability, and compliance with delivery times (low criticality, high availability) or the relationship between stability and the quality of the resources acquired (low criticality, low availability) count;
  - long-term: if the quality of the resources acquired also matters in the creation of competitive advantage (high criticality, high availability);
- *Business partners*: if vertical integration contributes to reducing risks or ensuring the current and future availability of particular resources (high criticality, low availability) [19];
- *Ecosystem partners*: organizations that belong to the global corporate network and that are weakly interconnected with each other, although dependent on each other for their survival [13];
- *Socio-cultural interpreters*: artists, cultural organizations, sociologists, anthropologists, marketing, and media experts who explore socio-cultural trends [13];
- *Technological process interpreters*: producers of innovative as well as professional machinery and plants that lead to the identification of technology trends in terms of production processes [13];
- *Technological product interpreters*: material as well as component manufacturers and professionals who contribute to identifying technology trends in terms of products [13];
- Technological connection interpreters: manufacturers of hardware and software components who contribute to identifying digital trends [13].

Procurement channels range from direct suppliers to physical points of sales where the material resources can be found or from industrial distributors up to digital procurement channels (B2B E-commerce, etc.). Finally, the inventory management methods include three different categories, as resources can be stored according to a 'fixed period' or a 'fixed quantity'; in addition, organizations rely on the just-in-time method in order to manage production and distribution, as goods are received by suppliers only as they are needed.

Resources

Resources are factors of production that can either be used once in the production process (raw materials, semi-finished, or goods) or for long periods within the company (land and buildings, plants and machinery). In addition to physical resources, a company may also have financial resources (cash, credit lines, etc.), which, together with the former, constitute tangible capital and intellectual resources. The latter can be divided into four categories:

- *Structural capital*, consisting of organizational procedures and routines, management systems, ICT, intellectual property, etc. [20];
- *Human capital*, consisting of the skills, level of education, experience, values, and social skills of the members of the organization [20];
- *Relational capital*, consisting of relations with customers, suppliers, complementary and competing companies, brands, image, etc. [20];
- *Environmental capital*, consisting of the historical, artistic, cultural, and landscape heritage characterizing the context in which the organization operates [21].

Resources are defined as distinctive if they are value-based, that is if they allow the company to respond to environmental threats and opportunities, rare, inimitable, and exploitable by the organization. Through the optimal allocation of resources, companies produce distinctive skills in order to distinguish themselves from competitors (e.g. miniaturization). Accordingly, knowledge is a particularly critical resource because it allows to combine resources and skills in new and distinctive ways, creating more customer value. Finally, resources are defined as *internal* if they are owned or directly controlled by the company and *external* if they involve relations with the environment and further corporate stakeholders [22, 23].

Processes and Outsourcing

Processes comprise the activities that the company develops in order to transform inputs (resources) into outputs (products). Considering their impact in creating a competitive advantage, they can be subdivided into the following categories:

- *Operational management processes*: procurement, production, distribution, and risk management of existing products [3];
- *Product development processes*: opportunity identification, R&D portfolio management, design, development, and product launch [3];
- *Market development processes*: selection, acquisition, loyalty, and further customer engagement strategies [3]; and
- *Social development processes*: activities aimed at environmental protection and social progress [3];

A company aims at developing different or similar activities as their competitors. This is also the reason why companies tend to focus on their core competences while outsourcing the development of less critical processes, even if these processes require high knowledge and skills, and therefore they are not merely operational [3, 9].

Products, Communication, and Distribution Channels

The products represent the visible offer which the company offers on the market for the purpose of fulfilling explicit (response), latent, or unidentified (proposal) customer needs. They can be subdivided according to the following dimensions:

- *Final destination*: products intended for consumption (B2C) or production (B2B);
- *Role in the offer*: loss leader, accessories, or complementors, etc.; and
- *Differentiation*: commodities (fungible and undifferentiated), goods (tangible and functionally differentiated), services (intangible and customized), and experiences (memorable and emotionally differentiated).

The latter can be divided into the following stages: co-design, purchase, delivery, use, integration, maintenance, and product disposal; it can also generate transformative experiences where the 'subject' to be transformed is the customer.

The communication channels are intended to increase the awareness and ability to evaluate the product. They can be divided into direct (postal channels, telephone, e-mail, etc.) and indirect (brand, product and institutional advertising, sponsorships, fairs, etc.). The distribution channels are designed in order to facilitate the product purchase as well as the product receipt and any assistance and after-sales service. Distribution channels can also be divided into direct (sales, e-commerce, etc.) and indirect (company-owned and franchised stores, distributors, etc.), therefore leading to multiple distribution strategies:

- *Intensive distribution*: placing products into as many retail locations as possible, without considering the characteristics and the quality of the stores. Suitable for distributing consumer products with low unit value, with frequent and impulse purchases for which the customer is not required to do any research before making a purchase [24];
- *Selective distribution*: placing products in more than one location, but not as many as with an intensive distribution strategy. Suitable for the distribution of complex consumer products (cars, telephones, etc.) and/or high unit value (jewelry, watches, etc.), or industrial products [24];
- *Exclusive distribution*: selling products through a specific storefront only. Suitable for products characterized by a strong brand image, for specialized and differentiated goods [24];
- *Direct distribution*: directly selling and sending products to consumers. The company can rely on door-to-door or online salesmen or on its own physical retail location, such as vending machines or stores. Single brand stores only in case of a vast range of products;
- *Immersive distribution*: positioning products in the market through stores or retail properties which, leveraging technologies like VR or AR or 360° images and videos as well as 3D visualization, can 'immerse' customers in brand values; and

- *Participatory distribution*: positioning products in the market through stores or retail properties which, leveraging the aforementioned technologies, not only can 'immerse' customers in brand values but can also encourage them to have an active role in the development of the corporate storytelling.

Customers and Customer Feedback

Customers represent the recipients of the products and customer feedback serves to determine whether or not the organization meets customer needs. In order to explain the buying motives of consumers, needs are hierarchically classified. Indeed, it is the satisfaction of lower order needs that makes higher order needs emerge. Needs can therefore be distinguished in:

1. *Physiological needs*: necessary for the survival of individuals, such as breathing, drinking, eating, sleeping, copulating, etc. These needs must be satisfied first due to the instinct of self-preservation and are therefore also found in new-borns [25];
2. *Needs relating to safety, tranquility, predictability*: absence of worries and anxieties on a physical, moral, professional level, etc. They arise when physiological needs are relatively satisfied and when infants become aware of the world around them [25];
3. *Needs relating to sense of belonging, friendship, affection, love, identification*, etc. arise when needs relating to safety are relatively satisfied and due to the growing aspiration of individuals, especially during adolescence and onwards, to belong to a social group [25];
4. *Needs relating to self-esteem, respect, approval, prestige, success*, etc. arise when the needs relating to the sense of belonging are relatively satisfied and due to the aspiration of individuals to feel competent and productive, especially after entering the workforce [25];
5. *Needs relating to knowledge, understanding, curiosity, exploration, learning*, etc. arise when the needs relating to self-esteem are relatively satisfied and due to the aspiration of individuals to become aware of the world around them [25];
6. Aesthetic needs and needs relating to harmony, beauty, creativity, etc. arise when the needs relating to knowledge are relatively satisfied and due to the aspiration of individuals to use their talent and knowledge for the purpose of creating and surrounding themselves with 'beautiful things' [25];
7. *Needs relating to self-fulfilment*, self-realization, and expectations, occupying a satisfactory position within a social group: arise when the needs relating to knowledge are relatively satisfied and due to the aspiration of individuals to become the best version of themselves [25];
8. *Needs relating to transcendence*, to overcome one's limits, to transcend all barriers of individuality by identifying oneself as part of a higher universe: arise when the needs of self-fulfillment are relatively satisfied and due to the acceptance of individuals of an altruistic, inclusive, or holistic vision of human consciousness [25].

The identification of the hierarchical level of the needs of the target customers contributes to building an effective value proposition. For example, if organizations want to attract customers whose safety needs have already been fulfilled, they must offer them functional products in order to make them feel part of a social group. In order to better satisfy customer needs, customers can be grouped into distinct segments with particular needs, buying behaviour, or further common attributes. A company must decide which segments to serve and which to ignore. It must decide, more specifically, whether to serve a:

- *Mass market*: not distinguishing customers based on their demographic, geographic, and behavioural characteristics, etc. The value offered, the distribution channels, and the market relations focus on one large group of customers with similar needs [26];
- *Diversified, segmented, niche, 'tailored' market*: clustering in order to identify various customer groups based on their demographic, geographic, and behavioural characteristics, etc. The value offered, the distribution channels, and the market relations can focus on similar or diversified segments depending on whether customer manifest needs, partially or totally different, to those of a single segment, of a niche market or of a single user [26];
- *Tribe*: clustering in order to identify various customer groups based on their passions, interests, and experiences that make them capable of collective action, since they are not mere consumers but rather supporters of the product. The value of the connection amongst members of a tribe is more important than the use value. Companies offering experiences can aspire to shift from serving market segments to serving customer tribes [27];
- *Multi-sided*: bringing together two or more distinct but interdependent customer groups through a digital platform. Such platforms generate value for one customer group, only if additional customer groups are also present. Platforms create value, facilitating interactions between the two groups.
- *Long tail*: distinguishing each customer from others, exploiting the possibilities offered by Industry 4.0 for the purpose of reducing the costs associated with custom production and distribution as well as the relationship between supply and demand. The focus shifts from selling high volumes of standardized products to selling low volumes of customized products [26].

Some customers contribute significantly to increasing margins and turnover; others are loyal over time and constitute valuable sources. In consolidated markets, customer feedback is functional to customer loyalty as it contributes to changing business processes for the purpose of creating products that best meet the needs of customers. In unexplored markets, on the other hand, it is functional to validate the hypotheses of (1) value—does the new product create value for someone? (ii) growth—does the new product have the potential to produce a significant growth?

Society and Corporate Social Responsibility

Society comprises the corporate stakeholders (employees, citizens, researchers, etc.) whose opinions and considerations might hinder companies from growth. In addition, society also constitutes the set of environmental, cultural, technological, economic, and political factors that characterize the external context. Society provides natural, historical, and artistic resources (e.g. the culture of beauty and the Made in Italy), but also human resources (e.g. interns) that constitute important competitive factors. Corporate social responsibility requires companies to exploit these resources and to deliver increased value to society. As a company turns a wood slab into a table, it must turn an intern into an expert.

However, by reading the business model canvas clockwise, social needs represent a fundamental stimulus to create new markets or businesses. The idea is to start from a social need, then (re)design products in order to fulfil it and imagine an innovative revenue model to make the business sustainable. Obviously, in order to create a new market, it is easier to start from an existing social need rather than to generate a new one [28].

The most notable example of strategic innovation arising from a social need is that of *Grameen Danone*: a joint venture between Grameen Bank Group and Danone Group (leading producer of yogurt). It was founded in 2006 as a social enterprise, involved in the production of yogurt, which is rich in important nutrients, for the children in Bangladesh who do not have enough food or proper nutrition. The company follows the 'no loss, no dividends' motto of social business, thus ensuring all profits are reinvested into the business for the purpose of ensuring sustainability, expansion, and growth. This is possible as raw materials, such as milk which is needed for production, can be sourced locally. Nevertheless, Danone Group was able to obtain significant returns in terms of:

1. Increased brand awareness as well as brand equity, in particular Grameen Bank was awarded the Nobel Peace Prize;
2. Entering a new and unexplored geographic market, where few milk-based products exist;
3. Reduction of production costs, due to the innovative technologies introduced as well as the R&D department; and
4. Change in the strategic vision of the company, which has shifted from 'a food and beverage conglomerate' to 'a family health and medical nutrition company'.

In other words, the corporate social responsibility actions can be grouped as follows:

- (Re)configuring operations in order to improve the social impact: a company may identify all societal needs that its value chain inevitably affects and is affected by (e.g. natural resources, total water use, working conditions, etc.) [29];

- (Re)designing products and creating markets in order to meet social needs: a company may identify all societal needs, benefits, and harms that are or could be included within the firm's products [29];
- Supporting the development of the (local) ecosystem: the success of every company is affected by the supporting companies and infrastructure around it (e.g. geographic concentration of companies, related businesses, suppliers, service providers, etc.) [29].

The adopted business model framework can therefore be read both clockwise, adopting an offer-based perspective, and counterclockwise, adopting a demand-based perspective. It presents three different levels of analysis:

- *Value chain*, that is the internal value chain of the company constituted by the functional resources needed to generate the products;
- *Value system*, that is the external value chain of the company which acts as an intermediate organization between its suppliers and its customers; and
- *Society*, that is the set of stakeholders, including those not belonging to the *value system*.

The adopted business model framework also presents four decision areas identified by several sub-triangles. The central one describes the aim to develop a new value proposition, the upper one identifies the aim to create a new market space, the lower left sub-triangle focuses on reducing unit costs and the lower right sub-triangle focuses on increasing unit revenue or selling price. The latter leads to the definition of the cost structure and the pricing model.

Cost Structure and Revenue Model

The cost structure considers both the cost variability (variable, semi-variable, fixed, and discretionary costs) as well as their origin/destination by considering the types of costs related to:

- *Production volume*: variable costs that increase or decrease depending on the production volume [30];
- *Production capacity*: fixed costs that depend on the installed production capacity [30];
- *R&D*: discretionary expenses aimed at developing new products and processes or modifying existing products or processes [30];
- Training and development: discretionary expenses generally included in indirect labour costs (supervision, quality control, inspection, purchases and receptions, material handling, etc.) [30];
- Marketing and communication: discretionary expenses necessary to market and distribute a product (employee compensation, advertising, storage, shipping, and customer service) [30]; and

- Relationship management: discretionary expenses necessary for maintaining relations between the company and the players within its microenvironment (suppliers, market intermediaries, public, and customers) [31].

The revenue model considers both the sources of revenues (sale, rent/leasing, pay-per-use, licensing, brokerage, advertising), and the pricing model. The latter is aimed at determining the sale price which, together with the product, promotion, and positioning, represents one of the major marketing levers [32]. Pricing models lead to several types of pricing:

- Economic, following the alignment with the lowest market price [32];
- Penetration, pricing products low enough to attract new customers [32];
- Premium, setting prices higher than the average market price [32];
- Loyalty, setting prices higher than the average market price, which can be reduced following subsequent purchases or bulk purchasing [32]; and
- Bundled, packaging several goods or services together as a single combined unit, often for a lower price than companies would charge customers to buy each item separately [32].

1.2.2 The Business Model Canvas and the Strategic Fit Matrix

The business model framework illustrated in Fig. 1.4 can be 'opened' in order to generate a canvas, thus representing a simple and immediate visual strategic tool. It is functional to map the business model of a company as it represents the first step of its redefinition. The areas directly attributable to the building blocks constituting the business model canvas are framed by two additional areas aimed at illustrating the impacts of the strategic choices in terms of cost structure and revenue model which constitute the profit model (Fig. 1.5).

In order to describe the function of the business model canvas, the Ikea case study is illustrated. Ikea has managed to revolutionize the furnishing sector, as in 2017 the business registered a positive result equal to 38.3 billion Euros, selling its products across its 403 stores located across 49 countries and counting 151,000 employees. The vision of Ingvar Kamprad, the founder of Ikea, was the following: 'To be on the side of the majority of people is a social ambition on which our business is based. The IKEA vision is to contribute to a better way of life for the majority of people. We do this by offering a wide range of home furnishings of good design and function, at prices so low that the majority of people can afford to buy them. That is our business idea. We know that in the future we may make a valuable contribution to the democratization process at home and abroad'. This thought was originated from the fact that in Sweden, during the late 1940s, younger households' needs, demanding new and low-cost furniture, were not satisfied as Swedish manufacturers only provided expensive items. This situation was viewed by Kamprad as both a social problem and a business opportunity, which he resolved by offering the majority of the population what they wanted: low-cost but high-value furniture. As stated in the

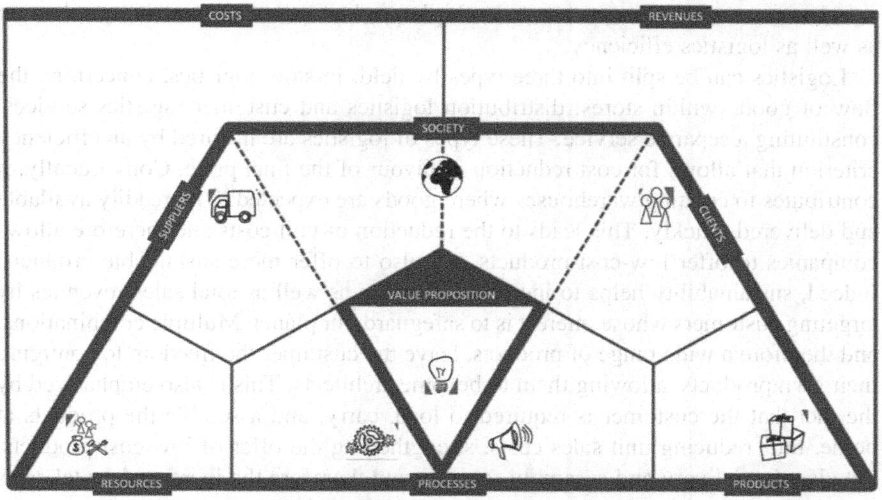

Fig. 1.5 The business model canvas

Testament of a Furniture Dealer, which resembles Kamprad's doctrine: 'All nations and societies in both the East and West spend a disproportionate amount of their resources on satisfying a minority of the population. In our line of business, for example far too many of the fine designs and new ideas are reserved for a small circle of the affluent. That situation has influenced the formulation of our objectives'.

By keeping in mind that fulfilling customer needs constitutes the ultimate source of a company's well-being in the long run, Ikea's scope then and today is 'to create a better everyday life for the many people', by offering a wide range of well-designed home products, at affordable prices for as many people as possible. As a matter of fact, Ikea offers a variety of products, although providing a basic range of essential items, therefore reflecting the company's mantra. Consequently, their focus is 'we don't just sell a chair or a table, we sell a philosophy and a mission', meaning that communicating their mission is as important as communicating their offering. Furthermore, as their offering is so essential and simple as their mindset and working environment, the company belief is 'less is more' [33, 34].

Their strategy, consisting in providing low cost, high-quality, and simple items, can be represented through the business model canvas, starting from the value proposition, 'we shall offer a wide range of well-designed, functional home furnishing products at prices so low that as many people as possible will be able to afford them'. In order to identify the business model of Ikea, we must consider that Ikea relies upon its product design team, therefore following an in-house approach, meaning no external designers are employed. Ikea's peculiar design is focused on essential aesthetic and modular architecture. This allows a reduction in design costs, but above all, it allows for the provision of functional furnishing products with clean and simple shapes, as an expression of Swedish culture. This also makes it possible to use common components across multiple products, thus maximizing production as well as logistics efficiency.

Logistics can be split into three types by field: in-store logistics, concerning the flow of goods within stores; distribution logistics and customer logistics services, constituting a separate service. These types of logistics are inspired by an efficiency criterion that allows for cost reduction in favour of the final price. Consequently, it contributes to creating warehouses where goods are expected to be readily available and delivered quickly. This leads to the reduction of unit costs and therefore allows companies to offer low-cost products, but also to offer more sustainable products. Indeed, sustainability helps to increase unit sales as well as total sales revenues by targeting customers whose interest is to safeguard our planet. Multiple combinations, and therefore a wide range of products, leave the customer the freedom to configure their own products, allowing them to become architects. This is also emphasized by the fact that the customer is required to load, carry, and assemble the products at home, thus reducing unit sales costs, strengthening the offer of low-cost products, but also the delivery and assembly services and therefore the increase in total sales revenues.

The company focused its efforts towards building long-lasting partnerships with both suppliers and customers. On the suppliers' side, Ikea relied on manufacturers coming from over 70 countries to build essential and modular products that comply with the unique design that distinguishes the Swedish company. Therefore, Ikea has no production facilities but designs all its products and exercises strict control over the network of sub-suppliers to ensure that they comply with the defined standards. As the design is developed in-house, maintaining relationships with suppliers is crucial. Ikea succeeded in this by establishing joint ventures with suppliers located in geographically isolated areas in order to set lower prices and ensure their control over rigid quality standards. This led to increased flexibility and reliability also from the suppliers' point of view, who can benefit from Ikea's expertise and knowledge to enhance their production efficiency. The profound evolution of the environmental context and the increased complexity of demand and competition have led Ikea to adopt a strategic orientation focused on the construction and consolidation of a network of relations with its stakeholders, especially when it comes to entering a new market.

On the customers' side, Ikea developed a system where the purchase experience required little or no assistance, as the products presented detailed instructions, and people were sent a yearly catalogue. Furthermore, the idea of self-service directly from the warehouse and self-assembly at home revolutionized customer relationships, fostering even lower costs from the manufacturing and shipping side. Ikea's motto, 'retail is detail', eventually summarizes the attention given to the customer experience while purchasing. The presence of warehouses characterized by real-time product availability allows the company to deliver goods quickly, and therefore providing instant gratification to customers. The stores, given their size and the wide parking area needed in order to facilitate furniture loading, are located outside the city centres and, more precisely, near motorway

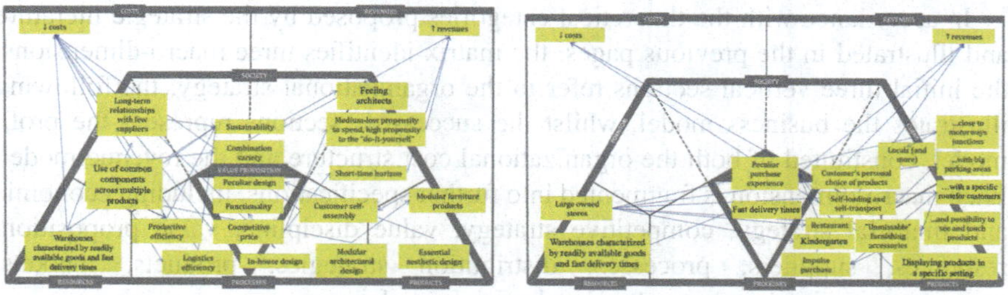

Fig. 1.6 The relationship among the elements constituting Ikea's business model

junctions in order to attract also non-local customers. The stores present some peculiarities (Fig. 1.6):

- Specific routes for customers to take through the stores, designed to 'force' them to visualize the entire offer and 'unmissable' products located at store checkouts;
- The possibility to look at and touch the products to create a visual, but also sensorial contact with the potential consumer, especially with children who are known to make impulsive decisions;
- Displaying products in a specific setting that facilitates the visualization of the context of use within the house;
- The presence of a restaurant in order to taste Swedish food and to appreciate Swedish culture; and
- Kindergarten located at store entrances to encourage couples with children to visit the stores in peace, therefore increasing their vulnerability to the products displayed.

The design of the products as well as that of the stores has allowed Ikea to target customers characterized by a medium-low propensity to spend, but a high propensity to the 'do-it-yourself' approach. Above all, the company was able to target customers characterized by a short-term horizon, therefore targeting a new market segment. Within this category, we can identify students who need to furnish their bedrooms during the study period or managers who are willing to change the furniture in their tied houses, as well as older adults with a lower life expectancy and who have no interest in purchasing products designed to last 20 years.

The identification of the business model of a company is functional not only to support the process of strategic innovation but also to evaluate, both ex ante and ex post, the degree of internal strategic coherence, that is alignment between the different building blocks constituting the business model canvas, and amongst these and the upstream business strategy, as well as the downstream profit model. To this end, it is useful to utilize the strategic fit matrix, which represents a tool used to visually verify whether the company is strategically aligned, and therefore coherent, in all its strategic dimensions characterizing it.

In accordance with the theoretical categories proposed by the strategic literature and illustrated in the previous pages, the matrix identifies three macro-dimensions: the initial three vertical sections refer to the organizational strategy, the following illustrates the business model, whilst the successive sections represent the profit model, constituted of both the organizational cost structure and the revenue model. Each macro-dimension is fragmented into further specifications, including economic and financial strategy, competitive strategy, value discipline, value proposition, resources, suppliers, processes, distribution strategies, products, markets, consumers, society, cost structure, and pricing model.

The horizontal rows highlight the strategic options to be pursued. The dimension of the value discipline, for example recognizes how the company can pursue: operational excellence, product leadership, customer intimacy, and ecosystem leadership strategies. The size of the product, however, recognizes how the latter can assume the qualification of commodity, good, service, experience, and transformation. For each of the strategic categories depicted, it is possible to recognize the options implemented within the organization: the green areas reflect options which have been implemented, the yellow areas refer to options which have been partially implemented, and the red areas represent the strategic options which have yet to be implemented.

Each dimension is also characterized by a strategic tension displaying two opposite poles. The degree of internal strategic coherence reaches its maximum once the strategic options highlighted in green appear all vertically aligned to the right side or to the left side of the matrix or anywhere else in between these two extremes. The strategic tension in the case of the discipline of value contrasts a closed strategy with an open one. In the first case, the company is focused on internal operations, while in the second case, the company is focused on external interactions, therefore involving as many companies as possible within the business ecosystem. Nonetheless, the strategic tension in the case of the size of the product contrasts with the offer of response to a proposal. In the first case, the company limits itself to responding to customer requests or proposals generated by competitors by offering a commodity, while in the second case, the company tries to develop innovative proposals by offering transformative experiences.

The matrix can thus be read both horizontally and vertically. In the first case, one dimension is analysed at a time in order to identify the theoretically coherent strategic option to be developed in order to increase the company's ability to create value. For example, if a company pursues operational excellence, it can theoretically evolve into product leadership and, only successively, into customer intimacy; just as if it offers a commodity, it can theoretically evolve into a good and, successively, into an experience. In the second case, that is reading the matrix vertically, the strategic options implemented (or to be implemented) in the different dimensions are compared to understand the overall strategic coherence. For example, the pursuit of operational excellence is theoretically consistent with the offer of a commodity, less with that of a good and theoretically inconsistent with the offer of an experience or a transformation. It follows that the degree of internal strategic coherence of a company is theoretically higher when the strategic options highlighted in green appear all vertically aligned to the right side or to the left side of the matrix, or anywhere else in between these two extremes.

The first case (strategic options aligned to the right side of the matrix) represents the situation emerging from the analysis of Apple which certainly pursues the economic-financial strategy of growth and the competitive strategy of domination, therefore pursuing ecosystem leadership. Apple proposes a new work and lifestyle, leveraging 'environmental resources', such as the IT engineers who graduated from Stanford University and who are based in Silicon Valley. Its suppliers are part of the ecosystem that it created, think about the App developers who use the App Store to market applications, and Apple also invests in environmental development processes since Apple's mission statement was, at the time of Steve Jobs: 'To make a contribution to the world by making tools for the mind that advance humankind' [35]. In consistence with the value proposition, it offers customers transformations that it also carries out by marketing its products through direct sales channels which are the Apple Stores. At the market level, it managed to create a tribe, the Apple fans, but the Apps are certainly attributable to a long-tail market. It satisfies customers' sense of belonging and self-esteem needs and supports the development of the local ecosystem which, as already specified, is that of Silicon Valley in California, which is not local, given the investments for opening the first iOS Developer Academy in Naples.

The second case (strategic options aligned to the left side of the matrix), represents the situation emerging from the analysis of Ikea, as shown in Fig. 1.7. The strategic fit matrix of Ikea allows us to notice that all its strategic dimensions are aligned to the left side of the matrix, with the exception of building a relationship with suppliers not in a pure market, where only price matters, but in medium-long-term partnerships. Ikea certainly pursues the economic-financial strategy of productivity, as well as of growth as shown by the economic-financial results previously discussed; as well as the competitive strategy of price leadership, and differentiation. Finally, it pursues operational excellence but, with reference to its price segment, it pursues product leadership. It offers a first price product, immediately available, without neglecting all the other critical success factors up to proposing a new lifestyle or, as Ikea itself points out, 'a philosophy of life'. To this end, it mainly leverages tangible capital, think about investments in store-warehouses and also structural capital, and think about the methods designed for visiting the store warehouses. It certainly utilizes technological interpreters of process, product and, albeit to a lesser extent, of connection, but also socio-cultural interpreters, starting from the Scandinavian designers. It invests in all types of processes, including those of environmental development: it was amongst the first companies in the world to embrace sustainability. It offers commodities and goods, as well as services and experiences, as previously explained. It focuses on the basic needs of customers (convenience, usability, functionality, and safety) and on aesthetic and creativity needs by offering design products. Consistently, it reconfigures its operations to reduce its environmental impact, but also redesigns products and markets to meet social needs, that is to allow everyone to use beautiful and well-designed items in their daily lives. From the analysis of the Ikea fit matrix, one can visually and,

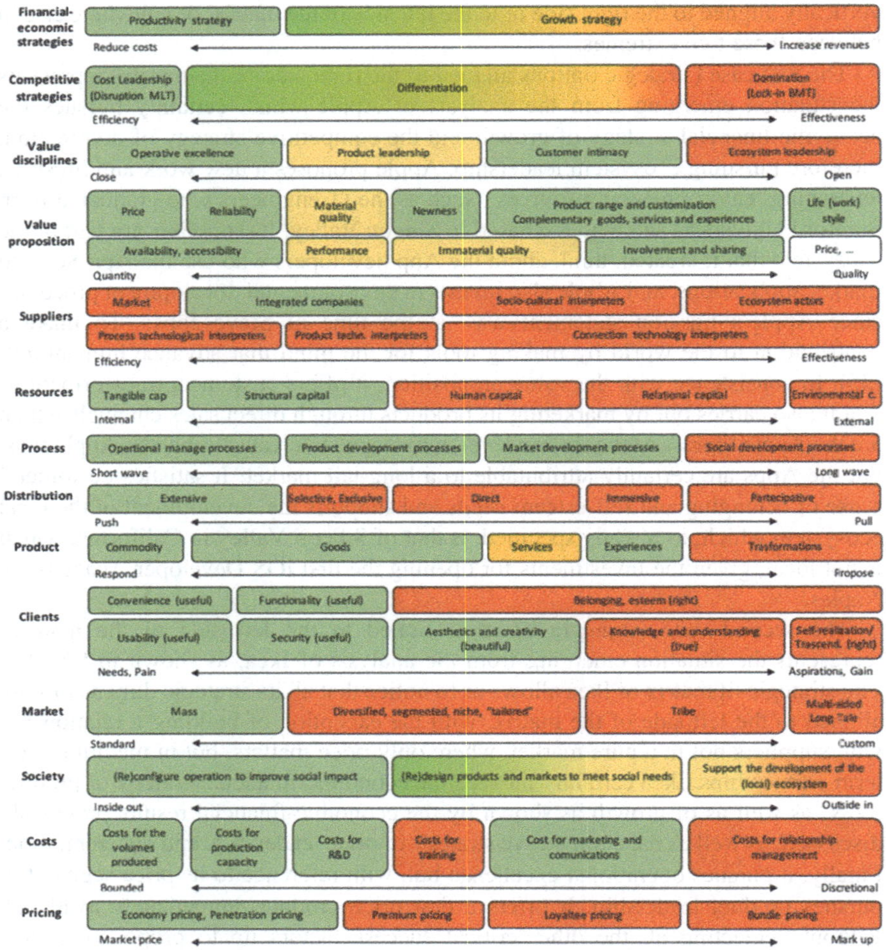

Fig. 1.7 Ikea's strategic fit matrix

therefore, easily notice the level of alignment amongst the strategic dimensions (reading the matrix vertically) and the potential challenges of each dimension (reading the matrix horizontally) in order to identify the theoretically coherent strategic option to be developed to increase the company's ability to create value. For example, if we consider the market size, one could try to create a tribe of Ikea customers, to fulfil their sense of belonging and self-esteem needs.

1.3 The Fourth Industrial Revolution

Industry 4.0 is the short expression used to identify the Fourth Industrial Revolution, namely the profound and irreversible digital transformation affecting the production system, as well as the socio-economic phenomenon, leading to exponential growth. To understand the scope of the present revolution, it is useful to recall the previous ones, highlighting the technological innovations characterizing them, as well as their impact on the business models of companies [36].

The First Industrial Revolution began in 1760 and mainly involved the textile, metallurgical, and extractive industries. It brought about radical changes within the socio-economic context, contributing to the development of trade and the rise of business, due to the invention of Watt's steam engine (utilizing coal to make the steam run the engine), powering the mechanical chassis developed by Cartwright, and the steam locomotive designed by Thompson. In order to exploit these technological innovations, the first factories were created and were characterized by a vertically integrated production and a bureaucratic organization for the purpose of increasing productivity. Centralized production was the characterizing element of the new business model and entailed a significant impact on production processes and tangible resources: new materials and machinery; but also, on structural and human resources: new organizational procedures and, therefore, new staff and operational tasks and skills.

The Second Industrial Revolution began in 1870 and affected the electrical, chemical, and oil industries. It was characterized by an unprecedented technological development, which was not the result of random inventions, instead it originated following scientific research carried out in university and non-university laboratories, financed both by governments as well as private entrepreneurs. These led to the invention of the petrol engine and of electricity which enabled Edison to create the first light bulb. In order to exploit these technological innovations, the division of activities carried out in the factories was established for the purpose of standardizing processes and products, as a matter of fact the moving assembly line was developed to assemble the Ford Model T. Mass production represented the characterizing element of the new business model and involved a significant impact on tangible and intangible resources and on production processes, as well as on R&D in order to design accessible products. In this regard, Ford's strategic vision became famous: 'a car in every garage'.

The Third Industrial Revolution began in 1970 and affected all manufacturing industries but was characterized by the rise of Information & Communication technology which brings together electronics, information technology, and telecommunications. The introduction of technological innovations such as semiconductors, personal computers, and the internet has radically changed the production systems making them more automated, and therefore less dependent on direct labour. Consequently, the economic sector became characterized by outsourcing opportunities and society became post-industrial. Flexible production was the defining element of the new business model and led to a radical change in terms of resources and processes, but also of products and supply chains. In fact, suppliers and customers have been more involved in R&D processes in order to pursue a more personalized production capable of satisfying the explicit needs of customers through increasingly differentiated products.

The Fourth Industrial Revolution is underway and will affect all manufacturing sectors by subjecting them to a digital transformation that will blur the boundaries of manufacturing. Compared to the previous revolutions, the current one will have an equal or higher impact on the production system, as well as on the socio-economic context, without requiring extensive replacement of industrial equipment. Smart reconfigurable manufacturing machines are going to be subject to regular updates, whilst final products are going to be equipped with surface sensors for the purpose of developing computational power and enabling network connectivity, in order to allow the development of the Internet of Things or the Internet of 'every-things' or things, services, data, and people or the Industrial Internet of Things. The latter will establish itself more slowly than the commercial Internet, due to the longer investments characterizing the B2B sectors. Compared to the previous revolutions, the current one will be characterized not by an improvement of the industrial equipment and the production systems of a single factory but rather by creating smart factories:

- Integrating its physical and virtual systems, considering the value chain (resources–processes–products), value systems (suppliers–company–customers) and, in general, the value ecosystems (a company made up of corporate stakeholders);
- Considering the entire life cycle of the product that starts from its design and its development to its disposal and its recycling phase.

The intelligent factory will be at the core of a data-driven business ecosystem: The essence of Industry 4.0 is to exploit data to generate new value [37–43] (Fig. 1.8).

1.3.1 Industry 4.0 Technological Strategies

The term Industry 4.0, which is the translation of the German expression Industrie 4.0, indicates the advent of the fourth digital revolution, thanks to the complete digitalization of production systems that allows to increase the horizontal, vertical, and end-to-end integration of business processes, making them more agile and responsive to changes. Compared to the previous revolutions, the current one is characterized by a more optimal utilization of material resources as well as human resources, starting from a better exploitation of the digital resources that make both products and intra and inter-processes intelligent, as a result of the creation of Cyber-Physical Systems (CPS) and Cyber-Physical Production Systems (CPPS) [44, 45].

Cyber-Physical Systems (CPS): consist of product systems (finished products, machines, or production plants) constituted of physical as well as virtual components. Physical components represent tangible devices equipped with sensors,

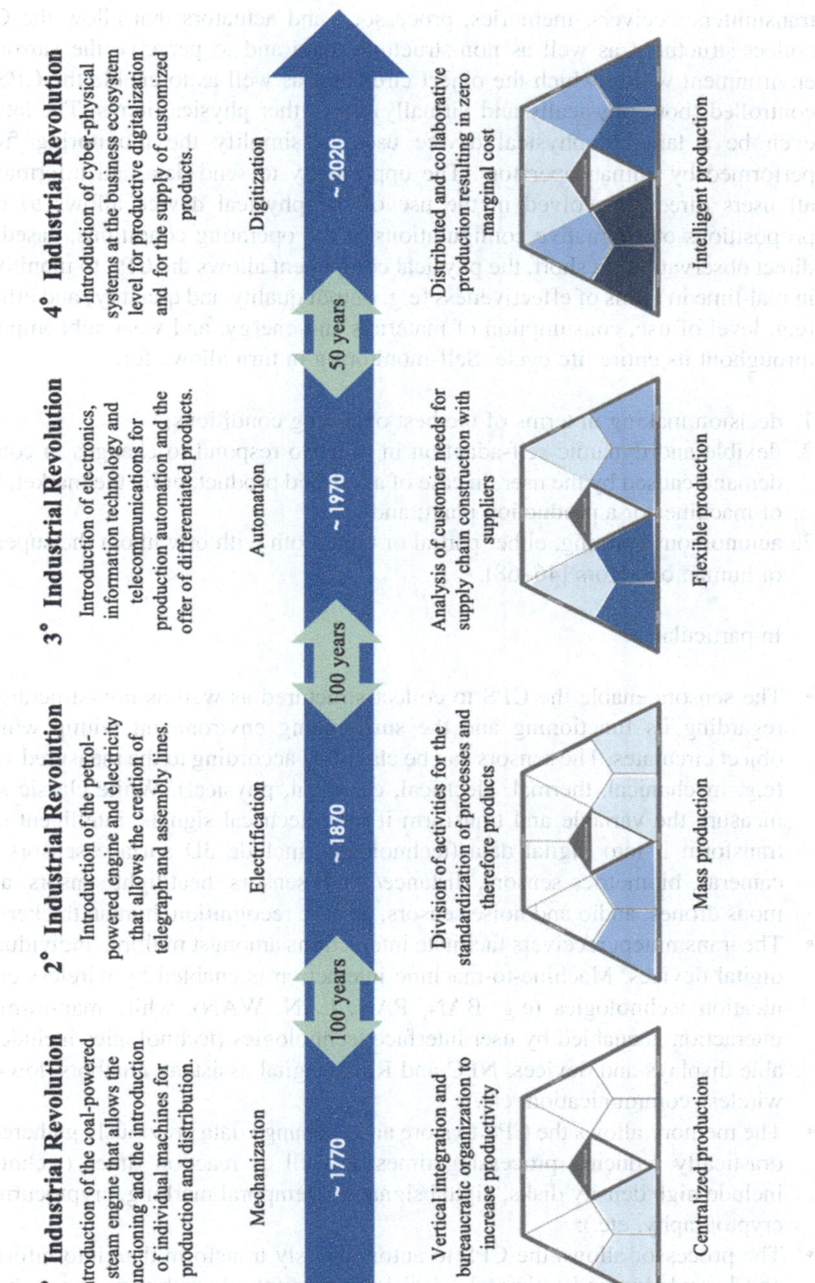

Fig. 1.8 Heatmap of the four industrial revolutions and their effects on the building blocks constituting the business model canvas

transmitters/receivers, memories, processors, and actuators that allow the CPS to collect structured as well as non-structured data and to perceive the surrounding environment within which the object circulates as well as to enable the CPS to be controlled (both physically and virtually) by further physical items. The latter can even be a tangible physical device used to simplify the monitoring function performed by human operators. The opportunity to send data and information to all users directly involved in the use of the physical device allows to receive propositions of alternative configurations of the operating conditions, based on its direct observation. In short, the physical component allows the CPS to monitor itself in real-time in terms of effectiveness (e.g. output quality and quantity) and efficiency (e.g. level of use, consumption of materials and energy, and wear subcomponents) throughout its entire life cycle. Self-monitoring in turn allows for:

1. decision making in terms of the best operating conditions;
2. flexible and dynamic self-adaption in order to respond to changes in consumer demand caused by the user, in case of a finished product, or by the market, in case of machines or a production plant; and
3. autonomous learning, either partial or total, both with or without the supervision of human operators [46–68].

In particular:

- The sensors enable the CPS to collect structured as well as non-structured data regarding its functioning and the surrounding environment within which the object circulates. The sensors can be classified according to the measured variable (e.g. mechanical, thermal, electrical, chemical, physical). While classic sensors measure the variable and transform it into electrical signals, intelligent sensors transform it into digital data (technologies include 3D surface sensors, smart cameras, biometrics sensors, distance/depth sensors, heat/light sensors, autonomous drones, audio and noise sensors, gesture recognition, human trackers, etc.);
- The transmitters/receivers facilitate interactions amongst multiple individuals and digital devices. Machine-to-machine interaction is enabled by wireless communication technologies (e.g. BAN, PAN, LAN, WAN), while man-to-machine interaction is enabled by user interface technologies (technologies include wearable displays and devices, NFC and RFid, digital assistant, chatbots, low-power wireless communication, etc.);
- The memory allows the CPS to store and exchange data previously gathered, thus drastically reducing processing times as well as reaction times (technologies include high density disks, digital signature, temporal marking, cryptocurrencies, cryptography, etc.);
- The processor allows the CPS to autonomously transform data into information and knowledge and ultimately decisions, through algorithm design techniques, therefore autonomously learning how to improve them based on feedback (technologies include neural networks, sentiment analysis, API management,

advanced CRM and ERP, natural language processing, business intelligence, embedded computing, etc.);

- The actuators enable the CPS to modify its operation in accordance with the decisions (both autonomously and non-autonomously) taken and to adapt to the surrounding environment (technologies include additive 3D printing, robotic manipulators, automatic access control, laser processing, speech synthesis, high efficiency batteries, autonomous driving, CNC machines, etc.).

On the other hand, the virtual component consists of a digital twin, defined as the digital copy of the physical device, simulating its operation during the design and utilization phases. The digital twin, during the product design process [69], allows to:

1. simulate its behaviour in order to prevent errors;
2. support its implementation by mirroring user requests; and
3. configure optimal operating conditions, thus limiting costs and risks.

While, during product utilization, it allows to [69]:

1. *Compare* real and expected performances, even simulating different scenarios;
2. *Reconfigure* the best operating conditions;
3. *Foresee* possible malfunctions allowing the development of predictive maintenance;
4. *Identify* reusable parts when disposing of the device [70]; and
5. *Support* its incremental or radical improvement as well as research and development activities, as a result of big data production [40], [71], [72].

The last functionality is probably the most important one. Digital twins collect data and information for the purpose of identifying products most suitable for consumers, thus enabling the pursuit of product leadership as well as customer intimacy strategies. In certain cases, it allows for data and information monetization, therefore producing additional revenue streams. The aforementioned functionalities are geared towards reducing operational costs, thus enabling the pursuit of operational excellence.

In short, CPS are *smart* system products starting from the product planning process and evolving through production and distribution, into its disposal. Furthermore, CPS are *adaptive* as they perceive the surrounding environment within which they operate, *customized* as they are able to self-configure the best operating conditions and learn from the way they are utilized and *pro-active* as they are able to self-diagnose and warn the technicians. They represent products with a low economic duration, but a greater physical duration, due to technological updates. They also represent products with a digital documentary evidence both in terms of function parameters and standards, allowing to share information for their optimization and further development and in terms of order batching, production, and assembly.

The functions of the digital twin echo those of Computer Aided Design (CAD), although the former also impacts on an organizational level leading to improved decision-making. CAD was introduced in the mid-1990s to support the digital design of a physical or virtual product. Product Data Management (PDM) appeared shortly after to support the management of the data created by the CAD, allowing better visibility of all phases of the product development process. Collaborative Product Commerce (CPC) appeared in the 2000s, which allows the sharing of resources between the company and its partners for the joint design of the product. The use of PDM and CDC then allowed the development of Product Lifecycle Management (PLM) which enables a strategic approach to manage the entire life cycle of the product, from its idea to its development, launch, and disposal. In the next paragraphs, we will consider the use of the digital twin and the digital thread [73–75].

Cyber-physical production systems (CPPS) constitute production systems consisting of multiple CPS [39] and further archiving systems, in addition to those already supplied by individual CPS and human operators. They are present along the entire value chain and value(eco)system and along the product life cycle, beginning with the customer order, the product planning process, its production, its distribution, and its utilization, as well as its disposal. They allow to share precise data and information in real-time (e.g. different stock levels, flawed pieces, changes relating to a specific customer order, demand level) to self-monitor, decide, adapt, and autonomously learn, both with or without requiring the supervision of human operators. The focus thus moves from the single machine or production plant to the overall production flow, both intra and inter company. Building a trustworthy provider–customer relationship represents a precondition for the sharing of data and information [51], as well as for the conflict resolution related to intellectual property protection [76].

CPPS represent the basis for the realization of the smart factory which allows for the real-time connection of:

- all business areas involved in the production process, enabling a production which is:
  1. automated, resulting in the reduction of manual labour jobs, human errors and direct labour costs, thus guaranteeing easier integration of low-skilled or older workers, as well as the creation of new, highly skilled job profiles, requiring continuous and advanced training;
  2. decentralized, allowing the reduction of transport, thereby further promoting sustainability (besides reducing consumption and scrap and recycling), concurrently reducing time-to-market, and increasing scalability;
  3. dematerialized, enabling multiple services to be delivered remotely;
- all value (eco)system actors, enabling a production which is:
  1. integrated, allowing the individual enterprise to align its business processes with those of further business partners, therefore enabling supply chain extension and interconnection with high quality services in order to manufacture 'hybrid' products and to create additional value networks [36]. An increasing

number of providers of logistic services, for example is shifting from simple order collection to pre-assembly services;
2. focused, enabling the individual firm to focus on core activities and outsource non-core activities to partner firms;
3. collaborative, allowing the single enterprise to diversify risks and to benefit from synergies in order to capitalize on business growth opportunities and to adapt to the volatility of markets, as well as to reduce the product life cycle;
4. flexible, enabling the set of partner firms to customize products within a series production, thus engaging consumers and business partners during the design process: the latter can even provide updated projects in order to realize them, while saving time and money [77], but also during the process of providing the material, thanks to the real-time availability of information regarding the customer [78].

The horizontal integration between the company and its providers, through the alignment of labour flows, enables, for example, the distribution of raw materials and semi-finished products anywhere, however such products can be identified anytime in terms of their evolution, present status (where and how) and alternative ways of further completing such products. On the other hand, horizontal integration between the company and its customers allows for customer loyalty through the augmentation of product performances as well as the provision of additional services thanks to the continuous collection of data and information along the product's entire life cycle (e.g. automatic spare parts delivery) [78]. Horizontal integration between the company, its providers, and its customers can generate new co-production models, and thus new business models [36].

CPPS are crucial for the creation of the digital thread. The digital thread constitutes the digital representation of the entire physical life cycle of products. In addition, the digital thread consists of the digital project, evolving through consumer co-design [79], into the digital control of manufacturing processes, in order to reduce human errors and improve quality, thus preventing production defects, resulting in the digital monitoring of product utilization, which allows to analyse its impact on customer productivity, or elimination. The digital thread enables product digitization and traceability, thereby sharing, integrating, and converting real-time data included within business information systems into information and knowledge, which serves to maximize operational excellence in terms of productivity (optimizing efficiency in materials and energy consumption and production time, as well as production capacity exploitation), quality of research and development activities, procurement, production, and sustainability.

The physical and digital convergence within the CPS and CPPS, namely the merging of hardware and software, currently characterizing the Fourth Industrial Revolution is favoured as a result of the multiple disruptive technologies and software innovations transforming entire industries and shifting businesses to digital firms and processes. Some are very innovative, while others are incremental improvements of existing technologies. However, technological progress is contributing to increasing performance and reducing costs, enabling the proper

Table 1.1 The 'generic' technological strategies and their conflicting objectives

Technology strategy aimed at increasing	
Openness and freedom of the system	Closure and control of the system
Decentralization	Centralization
Externalization	Internalization
Interoperability	Ownership
Usability	Security
Predictability	Reactivity
Augmentation	Automation
Flexibility	Efficiency
Complexification	Simplification

implementation of innovative technologies across organizations [80]. The paradigms that are shaping Industry 4.0 can be grouped as follows:

1. *Enabling software paradigms*: collecting, automatically transmitting (Industrial Internet of Things), and securely (trusted data management) storing (cloud computing) significant amounts of data and at relatively low price, given Moore's law;
2. *Fundamental software paradigms*: processing datasets (data-driven decisions) for the purpose of converting data into valuable information and knowledge in order to allow for automation based on knowledge, decisions, and learning, even that of machines and production plants (simulation);
3. *Integrative hardware paradigms*: promoting real-time human-to-machine communication (augmented and virtual reality) as well as machine-to-machine communication (horizontal and vertical integration), and also automated conversion from digital to physical (digital manufacturing).

It becomes critical for companies to understand which enabling technologies to invest in and how to integrate them to better seize the opportunities offered by Industry 4.0. Two 'generic' digital strategies, presenting conflicting operational objectives, have been identified in order to, with reference to production systems, increase their (Table 1.1):

- Openness and freedom to allow the achievement of an emerging balance.
- Closure and control to allow the achievement of a deliberate balance.

The 'generic' technological strategies are interpreted in terms of conflicting operational objectives since their antithetical nature drastically reduces the ambiguity of the interpretation. For example, the meaning of 'complexification' is more easily understood when it is stated that its opposite is 'simplification'. The conflicting operational objectives can then be understood as both dilemmas and trade-offs. A dilemma is a problem that allows for only two possible extreme solutions characterized by the choice of one of the two conflicting objectives. A trade-off, on the other hand, is a problem that allows for infinite possible intermediate solutions characterized by a different balance between the two conflicting objectives. Imagine, for example considering the tension 'Prediction vs. Reactivity' in designing a cobot that helps prevent human operators from getting injured in the workplace. An extreme theoretical solution may be the installation of a physical fence around the cobot that closes as soon as it starts working, creating a totally predictable context. The opposite extreme theoretical solution can be a cobot without any physical fence but equipped with sensors capable of blocking it before colliding with the human operator, creating a totally reactive behaviour. The real solutions will, however, be compromised. Consider, for example the contrast 'Usability vs. Security' in designing a system that protects a server from malicious attacks. An extreme theoretical solution may be to disconnect the server from the network, in which case the system would be totally secure. The extreme opposite theoretical solution can be the adoption of a totally open server, without passwords and whose services are exposed without requiring any authorization, in which case the system would be totally usable. A real balanced solution aimed at promoting usability over security in an ideal percentage equal to 70% can instead be a system that provides authentication with login and password. A real balanced solution, aimed at promoting security over usability, in a percentage always equal to 70%, can instead be a system that includes the use of smartcards.

In any case, the eight pairs of conflicting operational objectives characterizing the two 'generic' technological strategies can be summarized as follows:

1. *Decentralization* vs. *centralization*. Decentralizing refers to the process of distributing production activities, data, information, knowledge, as well as decisions across various operating units, either separated spatially and/or managerially. It involves, for example the use of cloud computing. On the other hand, centralization refers to the concentration of the aforementioned activities within few operating units, integrated on a spatial and managerial level. It involves, for example the creation of proprietary data centres [81];
2. *Externalization* vs. *internalization*. Outsourcing consists of a business practice through which organizations assign to third parties the completion of determined non-core activities. It is pursued, for example by resorting to Software-as-a-Service. Internalization, instead, refers to the procedures through which operations and processes are developed within the organization. It is pursued through the acquisition of skills and infrastructures not necessarily linked to the core activities of the company [82];
3. *Interoperability* vs. *ownership*. Interoperability enables the interaction and exchange of data and information amongst systems, regardless of their system architectures. It is pursued through the adoption of standard interfaces: common protocols and formats that define, respectively, what is transferred and how it is transferred. Ownership, on the other hand, limits the interaction and exchange of data, information, and knowledge amongst systems. It is pursued through the adoption of interfaces that are not freely accessible due to the lack of public technical specifications or the presence of property rights [83, 84];

4. *Usability* vs. *security*. Usability ensures that operating systems are easy to use. It is pursued by simplifying protocols and technical specifications (M2M usability) or human-to-machine interfaces (H2M usability). Alternatively, security guarantees that systems cannot be accessed or corrupted in the absence of authorization. It is pursued through the implementation of mechanisms in order to control access to data and through techniques, including cryptography, in order to verify their integrity, origin, and confidentiality (digital security) or, for example through the provision of physical barriers in order to prevent contacts between human operators and cobots (physical security) [85].
5. *Predictability* vs. *reactivity*. Predictability allows to detect in advance vulnerabilities or weaknesses within a determined operating system, eventually leading to system failure and to prevent and avoid occurrence of events potentially damaging its functioning. It is pursued through the formulation of deductive hypotheses or inductive extrapolations based on the comparison between historical and instantaneous data through statistical techniques (model fitting) or automatic learning (machine learning). Reactivity allows to identify software defects so as to prevent and repair failures within the system, solving software errors. It is pursued through real-time monitoring and the implementation of feedback mechanisms [86];
6. *Augmentation* vs. *automation*. Augmentation seeks to achieve cognitive and physical enhancements within human experience. By contrast, automation aims at introducing efficient and cost-effective machinery and processes for the purpose of minimizing human intervention and optimizing production. It is pursued by using unsupervised systems, capable of transforming data into information, knowledge, and ultimately into automatic decisions [73];
7. *Flexibility* vs. *efficiency*. Flexibility (or dynamic effectiveness) enables systems to seamlessly adapt to frequently changing production processes [87, 88]. It is pursued through modularization, low standardization, and formalization procedures. Alternatively, efficiency (or static efficiency) optimizes manufacturing processes, ultimately contributing to achieving greater productivity. It is pursued through high standardization and formalized integration procedures [89–91];
8. *Complexification* vs. *simplification*. Complexification refers to the process of increasing system functionality and performance. It is pursued by integrating new physical or digital components into the product or by increasing the number of operations necessary to carry out a process. Through simplification, instead, organizations intend to pursue complexity reduction strategies. This can be done by eliminating or integrating physical or digital components, replacing the former with the latter, or reducing the number of operations necessary to carry out a process [74].

For example, a software house can choose to adopt, with several variations, an open source or a closed source production system. This choice corresponds to specific orientations with respect to some of the eight aforementioned dilemmas. In particular, the development of software products according to:

- Open source, through decentralization and outsourcing practices, encouraging as many contributions as possible from both individual external developers and software houses interested in sharing the code. Nonetheless, most of the development will likely be done internally. Openness will encourage interoperability and usability in order to easily insert external contributions into the main code. This exposes it to more malicious cyber-attacks, although ensuring faster identification by being subject to multiple reviews by multiple developers. It will generally become more flexible as a result of its adaptation to several different cases.
- Closed source, through internalization, to avoid the uncontrolled disclosure of the code by unfaithful business partners. However, for what concerns large software houses, a certain degree of outsourcing is possible by subjecting the business partners to stringent NDAs. Closed source will favour ownership, although a certain degree of interoperability is possible through external APIs not directly connected to the source code. It will promote security, as the non-accessibility characterizing the code makes it more difficult for malicious hackers to identify its vulnerabilities. It can promote code efficiency by choosing monolithic, highly integrated solutions, the pursuit of which is possible by not having to share the development process.

The example proposed does not seem to concern the dilemmas: 'prediction vs. reactivity', 'augmentation vs. automation', and 'complexification vs. simplification'. Compared to the latter, it is interesting to observe how, if automation represents the third industrial revolution, augmentation could perhaps represent the fourth. Big data & analytics allow for the reduction of the dimensionality of data, while the virtual reality of navigating them in a multidimensional space makes production systems intelligible, despite their increased complexity. This makes it possible to return to human primacy. Finally, it is interesting to observe how, if the trend characterizing the previous industrial revolutions was to increase the functionalities and performance of systems, adding components (big & better strategy), the current revolution instead tends to break the systems into single micro systems capable of performing a single function and then recomposing them in different ways in order to solve complex tasks. This favours, amongst other things, the reutilization of software and hardware components. If simplification is about disjunction and reduction, complexity is based on distinction and conjunction. A special case appears to be that of additive manufacturing: the use of innovative materials and the 3D printing system allow for the creation of more complex products, thus simplifying the production process.

1.3.2 Industry 4.0 Strategic Innovations

The digital transformation of the manufacturing industry is enabled by the following nine digital paradigms:

1. Digital manufacturing
2. Autonomous and collaborative robots
3. Augmented and virtual reality
4. Industrial Internet of Things
5. Cloud computing
6. Simulation
7. Big data analytics
8. Cybersecurity
9. Structural horizontal and vertical system integration

These digital paradigms contribute to changing the way business is done by introducing advanced solutions that allow companies to redesign their value chain and re-interpret their role, therefore impacting the entire value (eco) system. It becomes critical for companies to understand which enabling technologies to invest in and how to integrate them in order to seize the opportunities offered by Industry 4.0. The present research identifies the potential impact of:

- Technological paradigms (e.g. 'Digital manufacturing') on the single building blocks of business models (e.g. 'Suppliers'), leading to the identification of specific Industry 4.0 operational benefits (e.g. 'digital manufacturing could reduce the number of suppliers, in particular those specialized in the production of complex parts, i.e., self-production of components');
- Operational benefits in pursuing a specific operational objective characterizing a 'generic' technological strategy (e.g. 'the reduction in the number of suppliers, in particular those specialized in the production of complex parts increases the *centralization*, *internalization*, *ownership*, *reactivity*, *flexibility*, *simplification* of the production system') aimed at increasing the closure and control of the production systems;
- Operational benefits in pursuing a specific critical success factor characterizing a value proposition, and thus a 'generic' competitive strategy (e.g. 'the reduction in the number of suppliers, in particular those specialized in the production of complex parts has a positive influence on *price*, *availability*, *accessibility*, *rapidity*') aimed at increasing operational excellence.

The second chapter explores the potential impacts aforementioned for each of the nine technological paradigms characterizing the Fourth Industrial Revolution. Below, we summarized how these impacts could make the value chain, the value system, and the value (eco)system smarter.

Smart Value Chain

Resources

The resources building block is influenced by all nine digital paradigms related to Industry 4.0. Indeed, the implementation of emerging technological paradigms can benefit various categories of resources within a company: financially, Industry 4.0

can help relocate costs and efforts to build efficient production sites or optimize the use of as-a-service infrastructures. However, a cost relief isn't the major advantage of exploiting new opportunities in technology: organizational procedures and routines and management systems, which are part of a company's structural capital, may also benefit from a reduction of items in stock, better transparency of the movement of goods throughout the supply chain and a more secure data storage. Human capital is also enhanced by the paradigms of Industry 4.0, by providing workers with the right tools to train and learn, thus feeling part of a growing society. Following this trend, relationships within and outside the company may be pivotal for creating more intimacy, and therefore brand identity. Exploiting technologies of Industry 4.0 within a company's value creation process may contribute to exploring the hidden environmental capital, paving the way to lower energy waste production processes.

Internal Processes

The internal processes building block is the most impacted area of the business model canvas. The technologies implemented within the operational management processes will enable the real-time planning of production, optimizing the work of human operators and machines thus reducing production times. By reducing time, they will also increase the speed of adaptation to different kinds of undesired events (e.g. production downtimes). Thanks to real-time monitoring of the production process, there will be a more productive environment where human operators and machines can work safely. Product development processes will also benefit from the advent of technologies characterizing Industry 4.0 by reducing prototyping times, thanks to virtual testing. Products will be monitored using sensors and that will foster the development of learning economies. Market development processes will be increasingly adapted in real-time to market dynamics, thanks to the use of descriptive analysis, which generate insights that help solve customer issues and accelerate the response. Finally, social development processes will enjoy a more responsive production process characterized by less waste and low emissions and focusing on labour rights.

External Processes

The external process building block is influenced by almost all nine digital paradigms characterizing Industry 4.0. New technologies establish stronger reliability on the distribution system in terms of costs and speed. The freedom of having a safe distribution route, a transparent tracking system, and a promising shift towards autonomous driving represent the pillars of the next evolution of a company's external processes. One of the major opportunities of implementing new technologies in this field can be a reduction of time-to-market and an optimization of just-in-time delivery. This will be possible, thanks to the use of descriptive analysis, which explores past customers' behaviours, and of predictive analysis, which identifies new strategies to attract customers by anticipating their needs. Assistance to the customer will be enhanced either via ensuring fast and reliable shipment of products or by offering online maintenance of smart products. However, close contact with the customer is also kept active through services that are provided remotely, based on the needs of individual customers. This allows them to obtain virtually guided 'do it yourself' assistance. This intimacy guarantees an even deeper knowledge of the customer and, consequently, their satisfaction.

Products

The products building block is influenced by many of the nine digital paradigms characterizing Industry 4.0. Emerging technologies contribute to creating customizable products with low effort in terms of maintenance, predicting when and how the product should be fixed. The technologies mark a progressive shift from a common, impersonal artifact to a unique, self-representing product with complementary customer-based services. The smart products of Industry 4.0 are equipped with new features and functionalities, for example they are able to exchange information on their use and their status with other smart devices and with the manufacturer to whom it indirectly provides information useful for its optimization and innovation. The software systems they are equipped with allow for their traceability and monitoring, ensuring improvements during the utilization phase. More often the line between product and service seems blurred: digital manufacturing increases the quality of the product, IoT enhances the possibility to associate automated and personalized services, and cloud computing offers the opportunity to connect seamlessly within a platform. The goal is to increase the interaction between a customer and the company through the offering.

Smart Value (Eco)System

Industry 4.0 also transforms the external value chain by reducing its complexity and allowing for a higher degree of end-to-end integration [32, 39, 92, 87–89] and automation [27, 32, 44, 93, 89–91], mostly thanks to digital technologies and CPS [32], which shift the physical boundaries of the enterprise, transforming it into an extended enterprise [32] where interactions with the industrial ecosystem take place in real-time [26, 27, 33, 82]. Interconnected collaborative platforms create a 'networked' production environment, based on flexible collaborations within the value system. Thanks to digital threads (product identification throughout the production process) [27, 94], supply flows are recreated on a virtual level, forming an integrated, seamless system [32]. This guarantees greater transparency and efficiency within the value system [27, 32, 33], above all by recognizing and resolving inefficiencies and risks at source and increasing the robustness and responsiveness of the value system to external changes [32]. The formation of a smart value system improves the ability to forecast as well as it enhances accuracy and contributes to achieving greater speed and collaboration within innovation processes [32]. Transformations take place at the level of:

Suppliers

The suppliers' building block is influenced by the following paradigms characterizing Industry 4.0: digital manufacturing, simulation, and structural horizontal and vertical integration. A different application of the same technology can lead to multi-purpose benefits in terms of achieving the most advantageous economic and social relations. Digital manufacturing can help to reshape a company's choice in terms of suppliers on the market, thanks to the possibility of self-producing complex components. This will lead to a reduction in the number of specialized suppliers. Also, business partners may be re-settled: the risk of ensuring future stable availability of resources may be no more critical, thanks to the possibility of foreseeing the costs in the choice of alternative resources and suppliers. Technological process interpreters may be more and more taken into consideration for the exploration of the next technological trends, and also technological connection interpreters will be part of a company's primary source of knowledge.

Customers

The clients' building block is influenced by more than half of the technological paradigms of Industry 4.0. A company's target client may enjoy a more intimate relationship with its product or service, thus increasing a sense of belonging due to the fact of having fulfilled his/her needs. Feeling part of a 'community' of creators who create their own offering will cause a stronger sense of affiliation which will eventually lead to the creation of brand ambassadors. Through the analysis of the customer's preferences, it is possible to identify their current unfulfilled needs and consequently shape their future desires. Technology will help companies build relationships based on reciprocal understanding: on the one hand, companies will understand their target market and help identify customers' desires, on the other hand, customers will explore and learn about the company, thus becoming aware of the world surrounding them.

Society

Stricto sensu, the society building block does not seem to be influenced by the nine technological paradigms characterizing Industry 4.0. However, *lato* sensu, Industry 4.0 transforms the business ecosystem by making it smart, distributed, adaptive, and open, with the characteristics of self-organization, scalability, sustainability, and relationship management typical of natural ecosystems [34]. Companies need to be part of an ecosystem to perform well and, therefore, develop partnerships and collaborations focused on customer needs, giving them greater flexibility [34–41]. Innovation activities can also occur between different companies, at an ecosystem level [32], as this manages to attract the best ideas from strategic partners quickly and cheaply, transforming them into innovations that generate competitive advantages for the companies to which they are transferred [98]. Within the smart ecosystem, community platforms facilitate relations between the various players, laying the foundations for sustainable innovations [32]. The latter can be of two types: outside-in innovations take place inside the company and are subsequently absorbed within the company; inside-out innovations are realized within the company and are transferred outside through partners and customers [32]. These innovation processes within the smart ecosystem are enabled by digital technologies, which facilitate knowledge transfers [32]. In addition to the benefits aimed at maximizing operational excellence in terms of productivity, quality, time, and sustainability, the enabling technologies of Industry 4.0 contribute to developing new business models [27]. In other words, they foster strategic innovations to create new market spaces through unique value propositions [99].

Value Proposition

Industry 4.0 opens significant cost reduction opportunities and, more specifically, it increases operational excellence by promoting process efficiency and improving productivity. It allows production to be planned based on actual demand trends, to establish supply and sub-supply relationships that enable real-time processing, to manage production processes as cyber-physical spaces, to carry out predictive maintenance on machinery and plants, to robotize, mainly by monitoring them remotely, logistics and warehousing systems, but also to enable customized production capacity on a large scale [100, 101]. Industry 4.0 makes it possible to significantly reduce the positive effects of economies of scale so that the small size of a company can become a source of advantage rather than a disadvantage: the size of the individual organization is no longer relevant but rather its ability to position itself at the most strategically appropriate node of its value (eco)system. Coherently, 42% of the 133 operational benefits identified by the literature review seem to have an impact on accessibility, availability, rapidity: 26% on price and 22% on reliability.

Industry 4.0 also opens great opportunities in terms of revenue-raising by enabling the pursuit of product leadership through product redesign in order to make it smarter and manage its entire life cycle up to its disposal. Coherently, 17% of the 133 operational benefits identified by the structured literature review seem to have an impact on performance; 13% on newness; 8% on immaterial quality, and 7% on material quality. Industry 4.0 also enables the pursuit of customer intimacy by encouraging the introduction of new pre-sales and after-sales services and the improvement of the ability to react quickly to market needs, thanks also to the possibility of having ever-closer physical and digital contact with the end customer, to the extent that the boundary between the supply of goods and services is increasingly blurred. Finally, Industry 4.0 enables the pursuit of ecosystem leadership by collecting, integrating, transforming, and sharing in real-time data contained in the various public and private information systems inside and outside the company. The potential value that can be unlocked from these 'data lakes' leads companies to reflect on possible ways to monetize them and undertake the necessary forms of collaboration with customers and suppliers to support the creation of a data-driven business ecosystem. Accordingly, 14% of the 133 operational benefits identified by the literature review referring to the framework appear to impact on complementary goods, services, and experiences; 8% on product range and personalization; 7% on engagement and sharing.

The greater potential impact of the Fourth Industrial Revolution on reducing costs rather than increasing revenues is probably due to an over-emphasis of interpreting Industry 4.0 solely from the perspective of technological innovation. Technological innovation at the level of individual building blocks must be accompanied by strategic business model innovation. In this case, the objective is to reduce costs by eliminating factors that are no longer critical to success, but at the same time increase revenues by introducing new critical success factors to design new value propositions in order to create new business models. Especially for SMEs, the real opportunity offered by Industry 4.0 is, in fact, the possibility not to recover operational efficiency to 'do the same things better', but to innovate their strategy to 'do different things or do the same things but in a different manner in order to stand out from competitors' [20]. The aim is not to 'play better than the others' but to 'change the rules of the game' [4], as large companies have set to make it difficult for SMEs to participate in the game. The industrial revolution underway opens great opportunities for SMEs to create new markets by identifying latent needs or creating non-existent needs by developing innovative business models. This is true at the level of the individual company and above all at the production chain level. By intercepting the technological and innovation drive linked to the ongoing industrial revolution, SMEs have the opportunity to exploit their potential to organize, integrate, and regulate the production chains to which they belong, moving from a fragmented model to an intelligent and interconnected ecosystem, in which companies work together to provide a competitive product. The aggregation of SMEs into networks of connected enterprises will also facilitate their access to financial and technological resources and, more generally, access to sources of knowledge.

To briefly summarize, the Fourth Industrial Revolution could enable a radical competitive repositioning of SMEs if the opportunities offered by the enabling technologies of Industry 4.0 are exploited to design new business models and to identify better the growing demand for customized products. The third chapter describes the business models characterizing Industry 4.0 identified by interpreting technology innovation also as a driver for strategic innovation. These business models have been grouped into four meta-business models 4.0: data-driven, platform, smart factory, and servitization.

In order to fully exploit the opportunities offered by Industry 4.0, many SMEs will have to invest in [32]:

1. the digital infrastructure in order to strengthen it in terms of computing capacity, connection, security, protection, and reliability given the amount of information that will be stored, transferred, and processed;
2. governance and leadership mechanisms to ensure a consistent implementation of digital transformation processes throughout the organization, avoiding the development of disconnected initiatives; and
3. people to develop new skills and ways of working within decentralized organizational structures and through virtual forms of collaboration.

Innovation benefits from collaboration amongst players. In particular, in cross-industry ecosystems while in the process of introducing complex smart manufacturing systems, it is unlikely that a single enterprise will be able to provide a complete solution. It is instead necessary to integrate resources and knowledge from different fields that might not belong to a single sector. Collaboration is introduced to generate value when it is difficult to create it from a single organization. Amongst the various forms of collaborative innovation, co-competition or cooperation represents the most important approach. The present concept merges the following terms cooperation and competition and indicates a model that companies utilize while innovating their business processes. Cooperative relationships can take the form of partnerships, alliances, or joint ventures and serve to reduce transaction costs as well as to increase know-how and manage resource dependence. Cooperation also promotes innovation, enables investments, and fosters the diffusion of digital technologies. This results in the creation of a value network enabling the identification and development of new business opportunities in the Fourth Industrial Revolution [102, 103].

Industry 4.0 contributed to transforming society over the past decades, and the driving force behind the aforementioned digital trends is referred to as exponential change, meaning that the so-called exponential organizations, which are involved in business computerization, have abandoned the logic of linear growth and have instead embraced the era of exponential acceleration, therefore producing large-scale transformations involving markets and further companies within all sectors. The current digital disruptions resulted in the emergence of a hypercompetitive business environment. The changing consumer preferences and market boundaries are forcing organizations to implement an advanced organizational structure in order to thrive in the digital era and drive meaningful change [103].

References

1. Bagnoli, C., & Biloslavo, R. (2012). L'Innovazione strategica nei piccoli e medi studi commercialistici triveneti. *Rivista dei dottori commercialisti, LXIII*(2), 247–273.
2. Lindgardt, Z., Reeves, M., Stalk, G., & Deimler, M. S. (2009, December). *Business Model Innovation: When the game gets tough, change the game*. The Boston Consulting Group.
3. Kaplan, R. S., & Norton, D. P. (2004). *Strategy maps: Converting intangible assets into tangible outcomes*. Harvard Business School.
4. Henderson, R. M., & Clark, K. B. (1990). Architectural innovation: the reconfiguration of existing product technologies and the failure of established firms. *Administrative Science Quarterly, 35*(1), 9–30.
5. Kim, W. C. (2015). *Blue ocean strategy, expanded edition: How to create uncontested market space and make the competition irrelevant.* Harvard Business School Press.
6. Angeli, F., Christensen, C., & Raynor, M. (1999). *Il dilemma dell'innovatore.* Harvard Business School Press.
7. Shaw, M. J., Shapiro, C., & Varian, H. R. (2000). Information rules: A strategic guide to the network economy. *The Academy of Management Review, 30*(2).
8. Lieberman, M. B., & Montgomery, D. B. (1988). First-mover advantages. *Strategic Management Journal, 9*, 41–58.
9. Porter, M. E. (1996). What is strategy? *Harvard Business Review, 74*(6), 61–78.
10. Gilmore, J., & Pine, J. (1998). Welcome to the experience economy. *Harvard Business Review, 76*(4), 97–105.
11. Hamel, G. (2004). *Leader della rivoluzione. Come prosperare in tempi turbolenti e fare dell'innovazione uno stile di vita.* Il Sole 24 Ore Libri.
12. Treacy, M., & Wiersema, F. (1993). Customer intimacy and other value disciplines. *Harvard Business Review, 71*(1), 84.

13. Verganti, R. (2009). *Design-driven innovation: cambiare le regole della competizione innovando radicalmente il significato dei prodotti e dei servizi*. Etlas.
14. Levy, S. J. (1959). Symbols for sale. *Harvard Business Review, 37*(4),117–124.
15. Bagnoli, C., Massaro, M., Dal Mas, F., & De Martini, M. (2018). Defining the concept of business model: A literature review. *International Journal Knowledge Systems Science*.
16. Hamel, G. (2001). Leading the revolution. *Strategy & Leadership, 29*(1), 4–10.
17. Anthony, S., & Schwartz, E. I. (2017, May). What the best transformational leaders do. *Harvard Business Review Digital Articles*, 2–9.
18. Iansiti, M. & Levien, R. (2004). *The keystone advantage: What the new dynamics of business ecosystems mean for strategy, innovation and sustainability*.
19. Kraljic, P. (1983). Purchasing must become supply management. *Harvard Business Review, 9–10*(5), 109–117.
20. Bontis, N. (1998). Intellectual capital: An exploratory study that develops measures and models. *Management Decisions, 36*(2), 63–76.
21. Claver-Cortés, E., López-Gamero, M. D., Molina-Azorín, J. F., & Zaragoza-Sáez, P. C. (2007). Intellectual and environmental capital. *Journal of Intellectual Capital, 8*(1), 171–182.
22. Braney, J. (1991). Firm resources and sustained competitive advantage. *Journal of Management, 17*(1), 99–120.
23. Nonaka, I., & Takeuchi, H. (1995). Knowledge-Creating Company. *Knowledge-Creating Company*, no. December 1991, 3–19.
24. Frazier, G. L. (1999). Organizing and managing channels of distribution. *Journal of Academic Marketing Science, 27*(2), 226–240.
25. Maslow, A. H. (2000). A theory of human motivation. *Psychology Review*, 1–87.
26. Movement, T., Osterwalder, A., Pigneur, Y., & Smith, A. (2010). *Business model generation*. Wiley-Blackwell.
27. Cova, B., & Cova, V. (2002). Tribal marketing: The tribalisation of society and its impact on the conduct of marketing. *European Journal of Marketing, 36*(5/6), 595–620.
28. Bagnoli, C., Biloslavo, R., & Edgarc, D. (2017). An eco-critical perspective on business models: The value triangle as an approach to closing the sustainability gap. *Journal of Cleaner Production*.
29. Porter, M. E., & Kramer, M. R. (2011). Creating shared value. *Harvard Business Review, 89*(1/2), 62–77.
30. Blocher, E. J., Cokins, G., & Stout, D. E. (2010). *Cost management a strategic emphasis*. McGraw-Hill/Irwin.
31. Ravald, A., & Grönroos, C. (1996, February). The value concept and relationship marketing. *European Journal of Marketing, 30*(2), 19–30.
32. Muller, G., & Nagle, T. T. (2015). *The strategy and tactics of pricing*. Routledge.
33. Kamprad, I. (2013). *The testament of a furniture dealer to create a better everyday life for the many people*.
34. Reichert, J. (1994). *IKEA and the natural step*. Oxford University Press.
35. Business Insider. (2013). *Apple's "Mission Statement"*. https://www.businessinsider.com/apples-new-mission-statement-2013-8?IR=T
36. Montanus, M., & Thesis, G. (2016). *Business models for Industry 4.0*. Delft University of Technology.
37. Bagnoli, C., Bravin, A., Massaro, M., & Vignotto, A. (2018). *Business model 4.0. I modelli di business vincenti per le imprese italiane nella quarta rivoluzione industriale*. Edizioni Ca' Foscari.
38. Berger, R. (2016). Industrie 4.0. *"How it reshuffles the economic, social and industrial mode"*.
39. GTAI. (2013). *Smart manufacturing for the future*.
40. Cotteleer, M. J., Mahto, M., & Sniderman, B. (2016). *Industry 4.0 and manufacturing ecosystems: Exploring the world of connected enterprises*. Deloitte University Press.
41. Grosvenor, R., Liu, Y., & Qin, J. (2016). A categorical framework of manufacturing for industry 4.0 and beyond. *Elsevier, 52*, 173–178.

42. Kaynak, O., & Yin, S. (2015). Big Data for modern industry: Challenges and trends. *Proceedings of the IEEE, 103*(2), 143–146.
43. Lee, M., et al. (2018). How to respond to the fourth industrial revolution, or the second information technology revolution? *Journal of Open Innovation Technology Market and Complexity, 4*(3).
44. Ardito, L., Garavelli, A. C., Panniello, U., & Petruzzelli, A. M. (2019). Towards Industry 4.0: Mapping digital technologies for supply chain management-marketing integration. *Business Process Management Journal, 25*(2), 323–346.
45. Lafferty, C. (2019). Sustainable Industry 4.0: Product decision-making information systems, data-driven innovation, and smart industrial value creation. *Journal of Self-Governance and Management. Economics, 7*(2), 19.
46. Buliga, O., Müller, J. M., & Voigt, K. I. (2018). Fortune favors the prepared: How SMEs approach business model innovations in Industry 4.0. *Technology Forecasting & Social Change, 132*, 2–17.
47. The Boston Consulting Group. (2016). *Digital in engineering and construction.*
48. C. F. I. (2015). *CFI, Roadmap per la ricerca e l'innovazione - Research and Innovation Roadmap.*
49. Franke, J. (2015). *A look at Industry 4. 0 – How the 4th industrial revolution could improve efficiency in the plastics industry.*
50. Berttram, P., & Schrauf, S. (2016). Industry 4.0: How digitization makes the supply chain more efficient, agile, and customer-focused. *Strategy &Technology.*
51. Ingwald, A., & Kans, M. (2016). Business model development towards service management 4.0. *Procedia CIRP, 47*, 489–494.
52. Barrientos, A., Crespo, E., Martín-Barrio, A., Peña-Tapia, E., & Roldán, J. J. (2019). A training system for Industry 4.0 operators in complex assemblies based on virtual reality and process mining. *Robotics Computer Integrated Manufacturing, 59*(July 2018), 305–316.
53. Ferraguti, F., et al. (2019). Augmented reality based approach for on-line quality assessment of polished surfaces. *Robotics Computer Integrated Manufacturing, 59*(April), 158–167.
54. Gadh, R., Lu, S. C. Y., & Shpitalni, M. (1999). Virtual and augmented reality technologies for product realization. *CIRP Annals – Manufacturing Technology, 48*(2), 471–495.
55. Baur, C. (2015). *Manufacturing's next act.* http://www.mckinsey.com/business-functions/operations/our-insights/manufacturings-next-act
56. Gilchrist, A. (2016). *Industry 4.0 The Industrial Internet of Things.* Apress.
57. Milan, F. (2016). *Bretonstone e l'Industry 4.0.*
58. Blanchet, M., Rinn, T., Thaden, G. von, & Thieulloy, G. de (2014). *The new industrial revolution: how Europe will succeed.*
59. Lu, Y., & Xu, X. (2018). Resource virtualization: A core technology for developing cyber-physical production systems. *Journal Manufacturing Systems, 47*, 128–140.
60. PricewaterhouseCoopers. (2015). *The Internet of Things: what it means for US manufacturing.*
61. Brettel, M., Friederichsen, N., Keller, M., & Rosenberg, M. (2014). How virtualization, decentralization and network building change the manufacturing landscape: An Industry 4.0 perspective. *International Journal of Mechanical, Aerospace, Industrial, Mechatronic and Manufacturing Engineering, 8*(1), 37–44.
62. Díaz-Garrido, E., Luz Martín-Peña, M., & Sánchez-López, J. M. (2018). The digitalization and servitization of manufacturing: A review on digital business models. *Strategic Change, 27*(2), 91–99.
63. Arnold, C., & Kiel, D. (2016). How Industry 4.0 changes business models in different manufacturing industries. *ISPIM Innovation Conference*, no. June.
64. Paine, T. (2015). *Industrial Internet of Things and communications at the edge.*
65. Anastasiadis, T., Lampropoulos, G., & Siakas, K. (2019). Internet of Things in the context of Industry 4.0: an overview. *International Journal Enterprise Knowledge, 7*(1), 4–19.
66. Raiwani, Y. P. (2013). Internet of Things: A new paradigm. *International Journal Scientific Research Publications, 3*(4).

67. Deloitte. (2016). Cyber risk in advanced manufacturing.
68. Burmeister, C., Luettgens, D., & Piller, F. T. (2015). Business model innovation for industrie 4.0: Why the 'industrial internet' mandates a new perspective on innovation.
69. Partners, S., et al. (2015, January). Industrial Internet of Things: Unleashing the potential of connected products and services committed to improving the state of the world. *World Economic Forum*.
70. Chryssolouris, G., Papacharalampopoulos, A., Petrides, D., & Stavropoulos, P. (2018). Dematerialisation of products and manufacturing-generated knowledge content: Relationship through paradigms. *International Journal Production Research, 56*(1–2), 86–96.
71. Manyika, J., et al. (2015). The Internet of Things: Mapping the value beyond the hype. *McKinsey Global Institute*, no. June, 144.
72. Heo, J. Y., Lim, C., Kim, K. H., Kim, K. J., Kim, M. J., & Maglio, P. P. (2018). From data to value: A nine-factor framework for data-based value creation in information-intensive services. *International Journal Information Management, 39*, 121–135.
73. Brandt, D., & Cernetic, J. (1998). *Human-centred approaches to control and information technology: European experiences organization technology.*
74. Tsoukas, H. (2017). Don't simplify, complexify: From disjunctive to conjunctive theorizing in organization and management studies. *Journal of Management Studies, 54*(2), 132–153.
75. Brant, K., & Halpern, M. (2002, October). *The differences among PDM, CPC and PLM matter.*
76. Koch, M., Merkofer, P. D., & Schlaepfer, R. C. D. (2014). Industry 4.0 challenges and solutions for the digital transformation and use of exponential technologies. *Deloitte.*
77. Gou, T. (2016, August). *Foxconn: Seeking diversification beyond contract manufacturing?* (pp. 1–11).
78. Bremicker, M., & Heynitz, H. v. (1990). *The factory of the future.*
79. Jirkovský, V., & Obitko, M. (2015). Big data semantics in industry 4.0. *Lecture Notes in Computer Science (including subseries Lecture Notes in Artificial Intelligence and Lecture Notes in Bioinformatics), 9266*, 217–229.
80. Hippmann, S., Landherr, M., Leis, M., & Neugebauer, R. (2016). Industrie 4.0 from the perspective of applied research. *Procedia CIRP, 57*, 2–7.
81. King, J. L. (1983). Centralized versus decentralized computing: Organizational considerations and management options. *ACM Computing Surveys, 15*(4), 319–349.
82. Nielsen, P. A., Nørbjerg, J., & Winkler, T. J. (2017). Closing the IT development-operations-gap: The devops knowledge sharing framework. *CEUR Workshop Proceedings*, volume 1898.
83. Hua, Z., Ke, W., Liu, H., & Wei, K. K. (2013). The impact of IT capabilities on firm performance: The mediating roles of absorptive capacity and supply chain agility. *Decision Support Systems, 54*(3), 1452–1462.
84. Appleyard, M. M., & Chesbrough, H. W. (2007). Open innovation and strategy. *California Review Management.*
85. Simota, J., Steiner, F., & Tupa, J. (2017, June). Aspects of risk management implementation for Industry 4.0. *Procedia Manufacturing, 11*, 1223–1230.
86. Bruton, K., Leahy, K., O'Donovan, P., & O'Sullivan, D. T. J. (2015). An industrial big data pipeline for data-driven analytics maintenance applications in large-scale smart manufacturing facilities. *Journal of Big Data, 2*(1), 1–26.
87. Collopy, F., Kwon, D., & Watts, S. (2002). Value frame, paradox and change: The constructive nature of information technology business value. *Working Papers on. Information Systems, 2*(17), 196–220.
88. Mouzas, S. (2006). Efficiency versus effectiveness in business networks. *Journal of Business Research, 59*(10–11), 1124–1132.
89. Adler, P. S., Goldoftas, B., & Levine, D. I. (1999). Flexibility versus efficiency? A case study of model changeovers in the toyota production system. *Organizatoin Science, 10*(1), 43–68.

90. Li, D., Patel, P. C., & Terjesen, S. (2012). Enhancing effects of manufacturing flexibility through operational absorptive capacity and operational ambidexterity. *Journal Operations Management, 30*(3), 201–220.
91. Ghemawat, P., & Ricart, J. E. (2017). The organizational tension between static and dynamic. *Strategic Management Journal, 14*(May), 59–73.
92. Alexandre, M. (2014). *Industrie 4.0 framework, challenges and perspectives*.
93. Erol, S., Schumacher, A., & Sihn, W. (2016). A maturity model for assessing industry 4.0 readiness and maturity of manufacturing enterprises. *Procedia CIRP, 52*, 161–166.
94. Hinterhuber, A., & Liozu, S. (2012). Is it time to rethink your ICT and education strategy? *MIT Sloan Management Review, 53*(4), 68–77.
95. Moore, J. F. (1993, June). *Predators and prey: A new ecology of competition*, 73–88.
96. Sveiby, K. E. (1997). The intangible assets monitor. *Journal of Human Resource Costing Accounting, 2*(1), 73–97.
97. Csik, M., Frankenberger, K., & Gassmann, O. (2016). *St. Gallen Business Model Navigator*.
98. Stock, T., & Seliger, G. (2016). Opportunities of sustainable manufacturing in Industry 4.0. *Procedia CIRP, 40*(Icc), 536–541.
99. Dosi, G., & Nelson, R. R. (2016). Technological paradigms and technological trajectories. *Palgrave Encyclopedia of Strategic Management, 1982*, 1–12.
100. Breunig, M., Kelly, R., Mathis, R., & Wee, D. (2015). *Industry 4.0-how to navigate digitization of the manufacturing sector*. McKinsey & Company.
101. Broberg, O., Kadir, B. A., & Souza da Conceição, C. (2019). Current research and future perspectives on human factors and ergonomics in Industry 4.0. *Computer & Industrial Engineering, 137*, 106004.
102. Siagri, R. (2021). *La servitizzazione. Per un futuro senza limiti di crescita*. Edizioni Angelo Guerini e Associati.
103. Vedovato, M., & Vaia, G. (2019). L'ambiente esterno. In C. Bagnoli (Ed.), *Economia aziendale tra tradizione e innovazione* (pp. 28–38). Giuffrè Francis Lefebvre.

2 Industry 4.0 and Its Digital Paradigms

The concept of technological paradigm was introduced by Dosi (1982), inspired by the work of Khun (1962). According to the latter, a paradigm is a set of elements able to solve a problem. Dosi defines a technological paradigm as a set of knowledge, such as scientific notions and research procedures, related to the birth and progressive development of a technology. A technological paradigm is therefore interpreted as a shared model, useful for solving problems that emerge during the normal progress of a given technology. Within the definition of technological paradigm, Dosi defined the term technological trajectory as the direction of advance within a technological paradigm, that is to say the pattern of "normal" problem-solving methodologies. In relation to this, incremental innovations represent a continuity along the technological trajectory of a paradigm, while radical innovations can lead to the emergence of new technological paradigms [1–3].

The technological paradigms of Industry 4.0 have the potential to innovate the business radically by changing completely customers' lifestyle, as the new pervasive digital environment is going to set new standards of interaction between humans and technology. Technological change at the business level enhances new organizational implications that eventually lead to the development of new business models. In order for this to happen, it is necessary to structure a business model that is aligned with the entire value chain. A clear business strategy is therefore essential to building a competitive advantage in the long term as it guides how to internalize and use the technological paradigms to remain competitive in the market.

Industry 4.0 will impact mainly four spheres: the supply chain, business, products and customers. Supply chain management will be more integrated and communicative due to the multiple ways to connect and control machines, resources and processes. Machines, in particular, will become intelligent and responsive to changes. Business is affected too, as Industry 4.0 will open new, stronger and more collaborative partnerships, optimizing demand-driven production in real time and access to complementary services. A totally new kind of product will be developed, and it will be more customized and digitized. Customers will be the

C. Bagnoli et al., *Digital Business Models for Industry 4.0*, Future of Business and Finance, https://doi.org/10.1007/978-3-030-97284-4_2

major actors taking part in the value creation process, requiring more additional functions and directly intervening within the design process [4–6].

The Fourth Industrial Revolution is marked by three fundamental blocks: paradigms, technologies and disciplines. While technologies are only a tool that the company can physically buy, paradigms and disciplines are more abstract elements: paradigms represent a strategic goal that the company intends to pursue, while disciplines represent knowledge of a given subject. The company that recognizes the new potential of Industry 4.0 in a manner consistent with its own identity can therefore identify the interventions necessary for its digital transformation.

As introduced before, a paradigm is a homogeneous methodological approach that makes some technological tools available to solve different issues, thus obtaining operational benefits. For instance, simulation is a paradigm that involves the use of models to solve, for example, optimization issues or fault prediction. The cloud is a paradigm that involves the use of remote and distributed systems to solve, for example, data security problems or remodulation based on the use of computational costs. Virtual reality is a paradigm that involves the use of viewers to solve, for example, issues related to telepresence.

A technology is a tool—hardware and/or software—that is used to solve problems in the context of a paradigm. A technology is implemented into a system according to the rules dictated by the discipline. Internet is a technology that enables the cloud through the rules, for example, of distributed computing and cybersecurity. Digital manufacturing is a technology that enables rapid prototyping—which is a paradigm—mediated by the discipline of robotics and materials science. Bayesian classifiers are a technology enabling sentiment analysis—which is a paradigm—through the disciplines, for example, of data science and social sciences. The technologies are divided into five classes: perception, communication, processing, actuation and storage.

A discipline is a set of knowledge and basic research and development capabilities that allow both the definition and adoption of paradigms and the design and use of technologies. The disciplines are transversal with respect to paradigms and technologies and are generally linked to the research and applied research activities of universities and public or private R&D laboratories. Data sciences and artificial intelligence (AI) are disciplines that allow, together with further disciplines, to develop analysis and prediction technologies, such as statistical interpolators and neural networks, and to adopt them within paradigms such as simulation or prediction. Some of the most known disciplines are artificial intelligence, data science, electronic engineering, mechanical engineering, computer vision, social sciences, materials sciences, cybersecurity, human-computer interaction and software engineering.

It is also important to clarify that paradigms, technologies and disciplines are subject to continuous changes dictated by innovation efforts. The evolution of these three concepts is completely different: disciplines vary by decades, paradigms vary by years and technologies undergo a very short life cycle and can be replaced by other technologies over the course of months.

The following paragraphs analyse the nine digital paradigms characterizing the Fourth Industrial Revolution:

1. Digital manufacturing
2. Autonomous and collaborative robots
3. Augmented and virtual reality
4. Industrial Internet of Things
5. Cloud computing
6. Simulation
7. Big data analytics
8. Cybersecurity
9. Structural horizontal and vertical system integration

The analysis starts with an initial description of each of the previously listed technological paradigms. Then, by analysing these technological paradigms through the lens of the business model canvas, we identify several "operational benefits", which represent the positive impacts that a company may obtain at the operational level by investing in some of the technological paradigms' underlying technologies. Finally, we graphically illustrate in a matrix the relationships between each operational benefit and (1) the building blocks constituting the business model canvas, (2) the different components constituting the value proposition and (3) the technological strategies.

2.1 Digital Manufacturing

Digital manufacturing practices have been adopted in a wide range of different applications since the early 1990s. However, it is only with the advent of Industry 4.0 that they finally became standard tools within the industrial and manufacturing industries. Still, the extent of such tools is not always fully understood by both managers and product executives. In fact, it is a common mistake to identify digital manufacturing with three-dimensional printing or other specific production and prototyping technologies. Any vertical technologies, including three-dimensional printing, laser cutting, digital milling machines or three-dimensional scanners, are just the physical incarnation of the digital manufacturing revolution. In a similar manner, an optical fibre can be regarded as part of the physical infrastructure of the Internet; still it is not the Internet, as the Internet is made up of protocols, applications and users.

Digital manufacturing, at least from an Industry 4.0 perspective, is not about being able to print objects on site. Digital manufacturing is about being able to melt the distance separating the design of an object from its physical synthesis. In other words, the core revolution is that, in a digital manufacturing world, once you have created a digital file describing the model of the object, you have created the object itself. This means that you do not have to worry about designing the production plant. Instead, you load the digital file into a black box, and you get the object just by adding time or money. Of course, such black box can be as simple as a three-dimensional printer, but more likely, for any non-trivial design, it will be an online service that in turn will use artificial intelligence, smart robots, dynamical supply chains and many other highly specialized Industry 4.0 tools to get the work done automatically and to deliver your item. This is extremely revolutionary, since the design of the production process is huge compared with the design of the product itself. Moreover, the design of the production process requires skills that are not always available for small business or freelance designer. Thus, being able to go straight from an idea to its synthesis is key to granting access to new markets to potential new players. Finally, the effortless translation of designs into physical objects changes forever the design process itself, allowing short test cycles and bridging the gap between the lean methodology and the physical world. For this to happen, three-dimensional printing technology is not enough. Actually, it does not even scratch the surface of the complexity of the problem. For this to work, further tools are needed, including file formats defining the physical and functional feature of the objects, algorithms to convert them in a sequence of operations, real-time reconfigurable machines and production lines, optimization of the supply chain, real-time procurement of raw materials, secure communication protocols and smart contracts to protect intellectual property. The good news is that in the last few years, many of these tools are becoming more and more readily available and the digital manufacturing ecosystem, including equipment manufacturers and service providers, is steadily growing.

The most prominent result of the application of these technologies in the manufacturing field is the need and opportunity to rethink the entire positioning of the company and the organization of the supply chain. In this context, digital manufacturing is a process defined by the technologies which enable the integration between product, process and resources. As discussed, digital manufacturing is defined as a set of digital technologies applied to the realization of goods or services and is therefore configured as the evolutionary step of manufacturing thanks to the incorporation of digital technologies.

Digital manufacturing is not defined by a single technology but rather by a complex bundle of technologies used in an integrated manner. The fact of being among the digital paradigms evokes the effort to aggregate all those information technologies that can somehow be used in the production system. The historical evolution of the traditional manufacturing model allows to see digital manufacturing as an opportunity to exploit more distributed and collaborative models, also driven by an ever greater involvement of the customer both in product design and its production.

Contrary to popular belief, adopting digital manufacturing technologies does not lead to abandoning traditional craft and manual manufacturing techniques; indeed, the result is a progressive convergence between traditional and modern models. Therefore, the fact of introducing production technologies into an artisan enterprise mutually benefits both spheres. Digital manufacturing becomes a tool to get closer to the specific needs of the customer, a typical element of craftsmanship; on the other hand, craft businesses can maintain their unique character and, by adopting technologies, can embrace new product development opportunities and increase production volumes. A company that adopts digital manufacturing technologies can benefit from innovative production processes that allow it to reach a market with increasingly complex needs.

With respect to the physical production, technologies empowering the digital manufacturing paradigm can be divided into two main areas: digital machining and three-dimensional printing.

Digital machining technologies comprise computer numeric control (CNC) machines, laser cutters and, in general, servo-operated tools. These technologies are part of subtractive manufacturing techniques and are closer to the traditional way of producing. These are just some examples of digital technologies used in the manufacturing system; the rate of change and technological evolution in recent years has reached exponential levels, causing two effects: a direct effect, given by the fact that only the most appreciated technologies "survive" over time, and an indirect effect due to the progressive decrease in investment in the technologies themselves. The fact that technologies are increasingly accessible is an opportunity for companies to embrace and implement technological progress to make their production models more flexible [7, 8].

Three-dimensional printing, also referred to as additive manufacturing, is a technology which is different from traditional technologies. Indeed, the subtractive process creates an object by removing material from a block; the forming process adds subsequent layers of raw materials until the final shape of the desired object is obtained. The process of additive manufacturing leads to the creation of an object using exactly the quantity and size of material it needs, without excess waste. This system works by "material addition", for it is also known as "layered manufacturing". Apart from cost reduction and improved flexibility, this method allows for rapid prototyping. First of all, rapid prototyping represents a series of industrial techniques in order to fabricate a prototype physically. With three-dimensional printing, rapid prototyping is achieved through simple steps: firstly, a digital model is created with CAD (computer-aided design) software; secondly, it has to be transformed into a STL (*Standard Triangulation Language To Layer*) file, which uses geometric data to decompose the area of the model into a series printable layers; and then, the file must be sent to a slicing software, which composes the slices for its printing; and finally, the slices are sent to the three-dimensional printer which builds the object layer by layer (Fig. 2.1).

Fig. 2.1 Rapid prototyping process

It has the potential of extending the production system's features and range of materials, allowing for continuous customization, as the three-dimensional model can be modified anytime. This eventually leads to mass customization. Although it will be extensively deployed in the upcoming years, the origin of additive manufacturing goes back to the 1980s, when Charles Hull invented the stereolithography to build three-dimensional objects. The difference from the modern three-dimensional printer developed by the Massachusetts Institute of Technology (MIT) is that now enterprises can manufacture products more rapidly and cheaply and without using toxic substances that were instead employed in stereolithography. Therefore, marginal production costs are lowered, and faster time-to-market is achieved.

As a matter of fact, the advantages of three-dimensional printing go beyond the organizational context, transforming the traditional view of factories. Once, large production facilities were built in remote locations, leading to a progressive distance between the production and consumption phases. Now, given the flexible and instant production enabled by three-dimensional printing, enterprises can relocate closer to their end users, closing the gap between producers and consumers and choosing strategic places in order to produce locally and connect with communities of three-dimensional developers. While making local manufacturing more profitable, additive manufacturing reduces the benefits of exploiting economies of scale. Nevertheless, it will be possible to create networks of collaborators. Additive manufacturing has the potential to transform designers into manufacturers, due to the accessibility of three-dimensional printing technology for both industrial and personal uses, at disposal of an active community that shares online designs. Indeed, the price of purchasing a three-dimensional printer has dramatically decreased in the last five years [9–11].

The most important benefit of introducing digital manufacturing into a business is the reduction in the number of **suppliers**, in particular those specialized in the production of complex parts thanks to the possibility of self-producing the components independently. This allows you to be able to meet market needs faster, while significantly reducing the need for financial resources and time to search for component suppliers. It is clear that digital manufacturing reduces purchasing and procurement management costs. This is because self-production capacity reduces supply chain dependencies, therefore reducing associated management costs [12–15].

The positive impact of the digital manufacturing paradigm on company **resources** is the centralized management of information and the reduction of unskilled labour; for this reason, products designed through additive manufacturing can be assembled quickly by even non-specialized personnel [13, 16, 17].

The use of digital manufacturing offers great advantages in **product creation processes**, first of all stimulating freedom in design and individual creativity, bringing out new ideas. The digitalization of the design enables a faster process of product development while maintaining quality and minimizing manufacturing defects. An example is three-dimensional printers and rapid prototyping.

Furthermore, the availability of virtual models facilitates the incremental modification of products; three-dimensional printing in particular allows for production on demand and its almost real-time realization. Mass customization enables the production of small batches thanks to the ability to quickly configure the machines to adapt to customer specifications. Three-dimensional printing allows also to print multiple materials simultaneously, thanks to the technique of adding materials in layers. The ability to produce different but integrated parts in one production step reduces the need for assembly activities. In addition to this, the possibility of building complex and already assembled geometries reduces the necessary components and increases the yield in materials [5, 10, 12–14, 18, 19, 16–24].

Even **external processes** can benefit from the advent of digital manufacturing. Indeed, the use of this technological paradigm enables access to remote or underdeveloped regions since technologies such as three-dimensional printing make personalized, decentralized and local production possible. Companies could print spare parts on demand in remote locations. Furthermore, it is possible to think of a new type of service, which allows the digital model of the product to be sent to the printing site closest to the customer, eliminating transport costs. This also benefits time-to-market and accelerates the ability to respond to changing market needs [10, 15, 18, 20, 23, 25, 26].

The **products** will be transformed by digital manufacturing: it will be possible to produce and ship small batches of customized products, much more complex and lighter. This occurs thanks to the progressive approach of production to the customer, for example, with three-dimensional printing. Adopting more flexible production lines will guarantee the design of more personalized and detailed products thanks to the integration of customer requests. Three-dimensional printing allows you to design new product features, adding complexity but without adding additional costs; moreover, it allows to reduce the level of inventory as well as delivery times [5, 12, 13, 16, 23, 25, 27–29].

Customers will also benefit from digital technologies, which increase collaboration opportunities and co-design, thanks to the constant communication between the company and its customers. Collaborative agility in real-time adaptation to changes in demand represents a key competitive advantage. The interaction with the customer will therefore allow greater control of expectations and even greater customer satisfaction [10, 16, 18, 27, 30].

Table 2.1 shows the operational benefits associated with digital manufacturing and the impact of these benefits on the building blocks constituting the business model canvas, on the technological strategies and on the value proposition.

2.2 Autonomous and Collaborative Robots

"Non-smart" autonomous robots are generally programmed to perform a repetitive operation, such as moving a piece, drilling it, screwing it, painting it and carrying out a sequence of fixed movements chosen from preset programmes. This happens without intelligent sensor-mediated management, except for security measures that

Table 2.1 The strategic impacts of digital manufacturing

Building blocks	Operational benefits	Technological strategies ("o" indicates a neutral impact)							
		Decentralization	Centralization	Externalization	Internalization	Interoperability	Ownership	Usability	Security
S	Reduces the number of suppliers, in particular those specialized in the production of complex parts (self-production of components)		X		X		X	o	
R	Reduces the use of unskilled labour and therefore the convenience of low-cost relocation strategies	o			X	o		o	
R	Reduces the optimal size of production sites (reduces economies of scale)	o		X		o		o	
R	Reduces the number of items in stock (elimination of components for products, self-production of some components)	o		o		o		o	
IP	Reduces the time and costs (including customs) of procurement (self-production of components)	o			X	o		o	
IP	Reduces product development time and costs (time-to-market and learning economies)	o		o		o		o	
IP	Reduces prototyping times and costs by digitizing the product realization process	o		o		o		o	
IP	Reduces the time and costs for the production of small batches (lead time eliminating physical set-up and reducing learning economies)	o		o		o		o	

IP	Reduces overproduction costs by adapting to market dynamics in real time	o	o	o	o
IP	Increases freedom in product design by fostering individual creativity and the birth of new ideas	o	o	o	o
IP	Enables the use of innovative materials	o	o	o	o
IP	Reduces waste and emissions	o	o	o	o
IP	Reduces assembly costs (fewer components)	o	o	o	o
IP	Increases the efficiency in the use of the material	o	o	o	o
EP	Increases the accessibility of the product even from remote areas	o	o	o	o
EP	Reduces the time and costs (transport, warehouse and customs duties) of delivering products to the customer	o	X	o	o
P	Increases the quality of the product (e.g. precision in processing, lightness of materials and performance)	o	o	o	o
P	Increases product customization	o	o	o	o
P	Enables the creation of more complex products (e.g. complex shapes and structures or different integrated materials)	o	o	o	o
P	Enables the creation of products with new features	o	o	o	o
C	Increases external collaboration in product design	X	X	X	o

(continued)

Table 2.1 (continued)

Technological strategies ("o" indicates a neutral impact)								
Prediction	Reactivity	Augmentation	Automation	Flexibility	Efficiency	Complexification	Simplification	Value proposition
	X	o		X			X	• Price • Availability, accessibility and rapidity
	X		X	X		X		• Reliability • Material quality • Performance
	X	o		X			X	• Availability, accessibility and rapidity
o		o			X		X	• Price • Availability, accessibility and rapidity
	X	o			X		X	• Price • Availability, accessibility and rapidity
	X	o		X		o		• Newness
	X	o		X			X	• Reliability • Newness
	X	o		X			X	• Price • Availability, accessibility and rapidity • Product range and customization
	X	o		X			X	• Price • Availability, accessibility and rapidity
	X	X		X		X		• Newness • Immaterial quality • Life (work) style
o		o		o		o		• Material quality • Performance • Newness • Immaterial quality
o		o		o		o		• Immaterial quality • Life (work) style
o		o			X		X	• Price
o		o			X		X	• Price
o			X	o		o		• Availability, accessibility and rapidity

	X	o		X		o		• Price • Availability, accessibility and rapidity
o		o		o		o		• Material quality • Performance • Newness
o		o		X		X		• Product range and customization
o			X	X		X		• Performance • Newness
o		o		o		o		• Performance • Newness
o		o		o		o		• Newness • Involvement and sharing • Product range and customization

prevent accidents, blocking the robot and then requiring human intervention. "Smart" collaborative robots make a combined use of sensors and artificial intelligence to give variable and context-appropriate answers following different environmental conditions or objectives. They differ from the traditional robots because they are able to think and act independently from the operator. Their cognitive power allows them to learn from mistakes and consequently avoid potential errors.

Their primary functioning process consists of working alongside with humans (*cobots*), that is to say, there are no more barriers between a human operator and a *cobot*, as the latter is designed to augment human labour rather than replacing it. They can basically do three things: perceive the environment and collect information about it; take consequent decisions, after having evaluated all possible scenarios; and finally, actuate in order to implement a specific task. Thanks to the Fourth Industrial Revolution, robots are embedded with sensors, which allow constant reception and transmission of data from the real world. Moreover, thanks to the deployment of artificial intelligence, collaborative robots will have more power to learn from their mistakes and change their algorithms to continuously adapt to new standards. This ability reflects the high level of flexibility that they can achieve during the production process. Indeed, robots can interact with each other and with humans, building a collaborative environment.

This includes some major advantages in terms of security and adaptability. For instance, *cobots* are able to see a piece that the human hands to them and take it even if it is not in a predetermined position or to respond to commands based on gestures. Sensors and artificial intelligence allow for autonomy, for example, by deciding how much force to exert with a drill based on the response of the material or to realign pieces before welding them together if they do not fit together perfectly. In essence, *cobots* are non-preprogrammed (or partially preprogrammed) and extremely adaptive collaborative robots. Their major advantages relate to human operators, as they preserve their security in the workplace, ensure quality products and increase productivity and thus competitiveness in the production process [8, 31, 32].

Autonomous and collaborative robots are considered a technological paradigm as they offer an innovative approach to redefine the industrial concept of automation through human-machine collaboration. This does not mean that robots will replace human work, rather they will cause a shift in the required human skills. Low-skilled human operators will no longer be needed, yet there will be an increasing need for augmented operators, who can interact with collaborative robots. As a paradigm, robots integrate different technologies like machine vision, geo-localization, haptic sensors, advanced mechatronics, cloud robotics and edge-enabled AI [5, 21, 33, 34].

The use of robots makes it possible to replace human resources in hard and stressful work activities, which will no longer be managed manually by workers themselves but by robots. Activities such as heavy-lifting, precision positioning and visual quality controlling will be replaced or supported by a safer and more effective technology than human operators, favouring cooperation between them in a common environment. Thus, the role of the simple worker is lost, and consequently, the figure of the skilled worker emerges, who performs few tasks but with a very high rate of responsibility and therefore of added value [5, 8, 21, 24, 33].

The impact of using autonomous and collaborative robots is predominant within **internal processes**. Automation in production planning and control thanks to robots makes it possible to process orders efficiently and without errors. This is because robots use data or information obtained from the machinery they work with or from the nearby production lines to make decisions and, if necessary, react to the unexpected. This happens thanks to advanced data mining systems and big data analysis. The introduction of robotic processes enables machine-to-machine communication, automatic learning and autonomous decision-making. This leads to improved productivity, precision and safety. An example of application of robots can be in support of assembly operations; robots could also be entrusted to the control of quality standards on the finished product, even in the case of production of small batches [4, 5, 7, 8, 14, 18, 20, 24, 25, 35–38].

As for the building block of **external processes**, the use of robots increases the use of more reliable transport systems, where transport carriers communicate their position and time to the smart warehouse management system. This system will lead to optimizing just-in-time and just-in-sequence delivery. This contributes to creating a potential opportunity to keep track of emissions and potentially save energy by optimizing delivery routes [39].

Table 2.2 shows the operational benefits associated with autonomous and collaborative robots and the impact of these benefits on the building blocks constituting the business model canvas, on the technological strategies and on the value proposition.

2.3 Augmented and Virtual Reality

Augmented reality offers an interactive experience in the real world, by enhancing or adding computer-based elements to it. While virtual reality creates a completely virtual world, isolating from the real one, augmented reality brings digital components into a person's perception of the real world, by integrating multiple sensory modalities, for instance, visual, auditory, olfactory, etc. It collects information based on human motion and translates it into augmented objects that can be seen through a variety of devices.

Augmented reality is configured in four different variants. The first one is called *marker-based*, and it works in the presence of a "marker" that is scanned by the camera, that is, a QR code. The second variant is called *marker-less*: it works via GPS and sends computer-based information about the location based on the user's current position. The third variant is called *projection-based* and works by projecting an augmented object, such as a hologram, onto a physical surface. The fourth variant is called *superimposition-based*, and it works by overlapping a physical object with an augmented object. The functioning of augmented reality is based on three steps: Firstly, input devices collect all relevant information that will be processed later. These devices can be either a smartphone's camera or sensors embedded into a *smart product*. Secondly, data are sent to a cloud server that uses a three-dimensional software to turn virtual images into augmented content to be then visualized by the user. Lastly, the visualization occurs with a display device that can be categorized in

Table 2.2 The strategic impacts of autonomous and collaborative robots

Building blocks	Operational benefits	Technological strategies ("o" indicates a neutral impact)							
		Decentralization	Centralization	Externalization	Internalization	Interoperability	Ownership	Usability	Security
R	Enables the replacement of human operators in repetitive, strenuous and risky activities	o		o		o		o	
IP	Increases efficiency and effectiveness in production planning and control	o		o		o		o	
IP	Increases flexibility without compromising the quality of the work	o		o		o		o	
IP	Enables complex processing by optimizing process variables (machine learning)	o		o		o		o	
IP	Enables complex processing by simplifying human activities	o		o		o		o	
IP	Increases labour productivity by promoting human-to-machine interaction	o		o		o		o	

IP	Increases the accuracy of the assembly simplifying human activities	o		o		o	o
IP	Enables decentralized production control	X		X		o	o
IP	Increases product and process monitoring using sensors (e.g. photocells)	o		o		o	o
EP	Increases the reliability of autonomous transport systems	X		X		o	o
EP	Increases the optimization of just-in-time and just-in-sequence deliveries	o		o		o	o
EP	Reduces emissions through autonomous and efficient logistic operations and delivery routes	o		o		o	o
P	Increases product customization	o		o		o	o

(continued)

Table 2.2 (continued)

Technological strategies ("o" indicates a neutral impact)								
Prediction	Reactivity	Augmentation	Automation	Flexibility	Efficiency	Complexification	Simplification	Value proposition
o			X	X		X		• Reliability • Material quality • Immaterial quality
o			X		X	o		• Price • Availability, accessibility and rapidity
	X		X	X		o		• Product range and customization
	X		X		X	X		• Reliability • Product range and customization
	X		X	X		X		• Reliability • Product range and customization
o		X			X	o		• Price • Availability, accessibility and rapidity
o			X		X		X	• Reliability
o			X	o		o		• Reliability
o			X	o		o		• Reliability • Material quality
o			X	X		o		• Availability, accessibility and rapidity
o		o		X		X		• Availability, accessibility and rapidity
o		o		o		o		• Availability, accessibility and rapidity • Immaterial quality
o		o		X		X		• Product range and customization

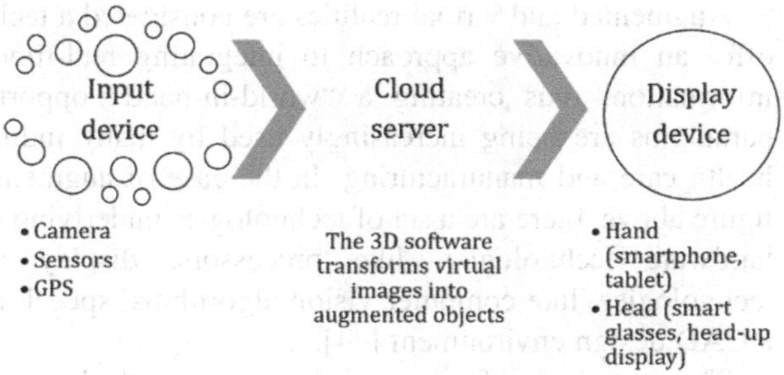

Fig. 2.2 The augmented and virtual reality ecosystem

a hand device (smartphone, tablet) or a head device (smart glasses, head-up display) (Fig. 2.2).

The history of augmented reality dates back to 1968, when Ivan Sutherland built the first head-mounted display (HMD) to recreate three-dimensional objects in the real world, but with limited capabilities. The term augmented reality was coined by Tom Caudell in 1990, and from this year onwards, this technology diffused and began to be used within the assembly lines and during machine maintenance. Since the 2000s, the first augmented reality devices appeared.

This technology may bring numerous benefits in the field of manufacturing, consumer decision-making process and human resource management, since actions taken by users have direct and immediate consequences on the surrounding environment. When applied in a factory, augmented reality can allocate information and expertise in situations where a remote maintenance support is required. Tracking labour and working movements in real time facilitates flexibility and labour productivity, apart from safety monitoring. Other applications of augmented reality in manufacturing involve the design-without-building and the optimization of assembly lines, due to the fact that it allows complete control over the system, by visualizing the internal machinery's components and its status and performance. Augmented reality can also enhance consumer decision-making, by configuring new or improved touchpoints, in which the company can interact better with its customers. In practice, augmented reality can help customers make better decisions during virtual shopping, enabling them to virtually try on the items. In fact, the shopping experience that leads to try an item online at home without physically going to the store adds value to the purchase choice. In the field of learning, this paradigm can lead to a radical change in the way processes are learnt. In fact, augmented reality offers a collaborative learning venue, in which many people, even physically distant, can learn together in a virtual environment in an immersive way. The greatest benefit lies in being able to use sensory modalities to communicate with people, to have a three-dimensional vision of the workplace and to personally take control and decide over the current situation. The increased learning experience can be applied to multiple areas, giving the opportunity to learn in complete safety and to connect people, instead of isolating them. Both augmented and virtual realities can be exploited in knowledge management and knowledge transfer. This makes it possible for a senior operator to share his/her expertise and competences with new operators [9, 34, 36, 40–43].

Augmented and virtual realities are considered a technological paradigm as they offer an innovative approach to integrating real-time environment with digital information, thus creating a "world-in-hand" opportunity for the user. These paradigms are being increasingly used by many industries such as construction, health care and manufacturing. In the case of augmented reality, as shown in the figure above, there are a set of technologies underlying this paradigm, ranging from hardware technologies—like processors, display and sensors—to software technologies, like computer vision algorithms, speech and gesture recognition and a CAD design environment [44].

The central part of a company's **resources**, the human workers, can benefit from a more controlled workplace, better learning resources and an excellent virtual space for training [40, 45].

The use of augmented and virtual reality predominantly impacts **internal processes**. A benefit lies in the location of products within the warehouse: wearable devices help the operator to view information on the exact position of an item on a shelf and to automatically scan barcodes. An additional advantage is the remote assistance for basic maintenance activities and the provision of product packaging instructions. The employees within the warehouse travel through optimized paths through the graphic display of the glasses, reducing the collection time of the right amount of the right article in a much more efficient and faster way. Thanks to this technology, employees can reduce training time for new employees. The biggest advantage at the industrial level is the ability to provide all of the information in real time in order to improve decision-making and work procedures, thanks to industrial assistance systems that use augmented reality to manage complex operations. By superimposing the relevant information on the employee's visual field, it is in fact possible to guide optimal production processes by reducing the associated risks [5, 18, 24, 28, 36, 39, 40, 45, 46].

For what concerns the building block dedicated to **external processes**, we expect a strong change in the way the products will be exposed to the customer. It will be possible to make aesthetic and functional tests on the product, measuring its performance and compliance with the customer before buying it. This eventually allows the customer specific needs to be adapted. Thanks to wearable devices, it is also possible to guide the consumer through product maintenance procedures without the need for the physical presence of a technician. The glasses in fact allow the technician to visualize the three-dimensional overlays on the physical objects to see how to remove or replace the parts and suggest the appropriate actions to the client. The result is a reduction in repair times, a damage reduction by unskilled technicians and a punctual support [15, 24, 40, 41, 47–52].

Following the introduction of augmented and virtual reality, new interaction mechanisms with **customers** are enabled through new or improved contact points, contributing to understanding customer needs and consequently increasing the customer experience [24, 27, 33, 37, 53, 54].

Table 2.3 shows the operational benefits associated with augmented and virtual reality and the impact of these benefits on the building blocks constituting the business model canvas, on the technological strategies and on the value proposition.

Table 2.3 The strategic impacts of augmented and virtual reality

Building blocks	Operational benefits	Technological strategies ("o" indicates a neutral impact)							
		Decentralization	Centralization	Externalization	Internalization	Interoperability	Ownership	Usability	Security
R	Enables a safer and more controlled training environment for workers	o		o		o		o	
R	Reduces training time for new and seasonal workers	o		o		o		o	
IP	Increases the accuracy in the location of the goods in stock	o		o		o		o	
IP	Reduces the search time of the goods in the warehouse identifying the most efficient collection path	o		o		o		o	
IP	Reduces the time for unloading the goods from the warehouse thanks to the navigation instructions and the scanning of the bar code	o		o		o		o	

(continued)

Table 2.3 (continued)

IP	Increases the conformity of the goods taken from the warehouse with what is requested	o	o	o	o
IP	Increases the speed of decision-making and work procedures by providing workers with real-time information	o	o	o	o
IP	Enables new forms of human-to-machine interaction in production and maintenance processes	o	o	o	o
IP	Increases quality control	o	o	o	o
IP	Reduces human error and downtime by increasing the availability of all relevant information in real time	o	o	o	o

IP	Increases collaboration and learning on the field among more people remotely in production and maintenance processes	X		X		o	o
IP	Enables the virtual testing of product concepts in order to better understand customer needs	X		X		o	o
EP	Reduces repair times through guided maintenance procedures	o		o		o	o
EP	Enables the possibility of making aesthetic and functional product tests with customers	X		X		o	o
P	Enables the expansion of the after-sales service and the development of new services	X		X		o	o
C	Increases the customer experience	o		o		o	o

(continued)

Table 2.3 (continued)

Technological strategies ("o" indicates a neutral impact)								
Prediction	Reactivity	Augmentation	Automation	Flexibility	Efficiency	Complexification	Simplification	Value proposition
o		X		o		o		• Reliability • Immaterial quality
o		o			X	o		• Price • Availability, accessibility and rapidity
o			X		X	o		• Availability, accessibility and rapidity
o			X		X	o		• Availability, accessibility and rapidity
o			X		X	o		• Availability, accessibility and rapidity
o			X		X	o		• Reliability • Availability, accessibility and rapidity
	X		X		X	o		• Availability, accessibility and rapidity
	X		X	X		o		• Reliability • Availability, accessibility and rapidity
o			X	o		o		• Reliability • Material quality
X			X		X	o		• Reliability • Availability, accessibility and rapidity
	X	o		o		o		• Reliability • Availability, accessibility and rapidity
o		o		o		o		• Performance • Newness • Involvement and sharing • Product range and customization
	X		X	o		o		• Performance
o		o		o		o		• Involvement and sharing
o		o		o		o		• Complementary goods, services and experiences
o		o		o		o		• Complementary goods, services and experiences

2.4 Industrial Internet of Things

Industrial Internet of Things (IIoT) integrates information and communication technologies (ICT) within the manufacturing system, connecting devices of different nature by embedding sensors in order to communicate with each other as well as with a centralized unit of control.

In order to implement Industrial Internet of Things technologies, it is important to select the "things" to be identified via the Internet; in addition, data must be collected and managed within an IIoT platform and integrated in a cloud server. Indeed, if Industrial Internet of Things represents the intelligence at the edge of the system, the cloud can be viewed as the central brain to which every information is integrated. Every source of data collected from an individual machine or product is directed towards a central data storage. Sensors can be connected not only to the machines but also directly to the final products that will be distributed to the consumer, allowing the system to constantly receive information on its use and therefore to think faster and more efficiently in order to propose customized solutions (Fig. 2.3).

The advantage of working with an IIoT and cloud computing system relies in real-time access to resources, avoiding redundant information by storing data in one place and ensuring autonomous and robust connections among devices and applications. By connecting all the "things", it is possible to have a complete vision of what is happening to every part of the system. The possibility of collecting data even from assets that are physically distant through the use of sensors and a network favours the automation of the system itself, which makes it possible to achieve optimal production with minimum machine downtime. Moreover, with IIoT, workers will be able to manage their area with more flexibility, without the need for an on-site expert, as there will be real-time connection to the central command unit. Hence, workers will be able to concentrate more on their core activities, and there will be major savings in the employment of certified personnel to assist in the supply chain management. With a more digitally integrated supply chain, there will

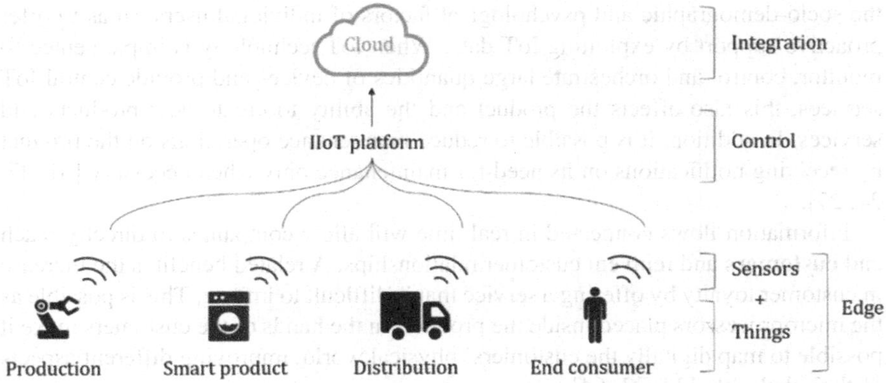

Fig. 2.3 The IIoT infrastructure components

also be significant energy savings that will lead the company to be more sustainable [5, 34, 55–58].

Industrial Internet of Things is considered a technological paradigm as it offers an innovative approach to turn data into relevant resources, by combining "things" with sensors. The smart technological ecosystem consisting of various interconnected devices will increase the transparency of the production system and make more precise decisions based on real-time data. As described in Fig. 2.3, the paradigm of the IIoT is enabled by a number of technologies, like hardware technologies, devices like RFID (radio frequency identification) sensors, GPS (Global Positioning System) chips, etc.; analytics solution technologies, edge computing to perform local data preprocessing or local data storage; and software technologies—machine-to-machine communication, the cloud and cybersecurity [59].

The use of IIoT systems in **resource** management favours the exchange of information and the control of actions. The availability of technical data of the components thanks to the real-time location of materials and products along the supply chain favours the creation of a constant updated knowledge [4, 24, 60].

The benefits on **internal processes** are manifold. With the industrial Internet of Things, a greater number of devices will be enriched with sensors equipped with integrated processing, allowing devices to communicate and interact with each other and with a centralized server. The so-called edge computing allows to decentralize some minor pre-analysis processes, thus avoiding overloading the server with information and allowing answers to be elaborated in real time. Logistics as well as the entire production process become more efficient, thanks to more detailed and up-to-date information that allows operators to monitor the condition of the machinery. As a result, the number of manual inspections and unexpected failures can be drastically reduced, along with maintenance costs [5, 24, 33, 36, 61, 62].

In relation to the **external processes**, the use of IoT technologies allows the monitoring of the conditions of transport of packages to customers. By adopting sensor technology, intelligent products will allow companies to access customers more easily to facilitate better after-sales services [24, 61, 63].

The **products** thus become the basis for identifying the opinions of customers and the socio-demographic and psychological factors of individual users, so as to offer proactive support by exploiting IoT data. When IoT technology is implemented to monitor, control and orchestrate large quantities of devices and provide central IoT services, this also affects the product and the ability to create new products and services. In addition, it is possible to reduce maintenance operations on the product by receiving notifications on its need for maintenance only when necessary [61, 47, 24, 27].

Information flows connected in real time will allow companies to directly reach end **customers** and reinvent customer relationships. A related benefit is the increase in customer loyalty by offering a service that is difficult to imitate. This is possible as the microprocessors placed inside the products in the hands of the customers make it possible to map digitally the customers' physical world, improving different aspects of their daily life [24, 33, 64].

Table 2.4 shows the operational benefits associated with Industrial Internet of Things and the impact of these benefits on the building blocks constituting the business model canvas, on the technological strategies and on the value proposition.

2.5 Cloud Computing

Creating a data-driven economy also requires increasing channel capacity. In this regard, Gilder's law establishes that "total bandwidth of communication systems triples every 12 months", thus translating into the ability to transmit large amounts of data at the speed of light. In the absence of exponential bandwidth growth, the concept of cloud computing would have never been formulated [62].

Cloud computing is a virtualized infrastructure that allows the use of online computing, storage, database, network, analysis and business intelligence resources and more. Through the Internet, it is possible to process all the organizational data into a central online hub, paying per use and thus saving on operational costs. Having access to shared resources anytime, anywhere and from every device represents the major benefit of exploiting an online infrastructure, which ensures efficiency, flexibility and scalability. Three essential elements constitute the cloud: a front end, comprising the client's access to the cloud; a back end, constituted of the provider's cloud infrastructure; and the network, connecting all related parties.

From the front-end user's point of view, he/she can have access to three different services: software as a service (SaaS), platform as a service (PaaS) and infrastructure as a service (IaaS). Infrastructure as a service (IaaS) provides customers with access to storage, network connection, servers and additional computing resources in the cloud. The virtualization of administrative tasks allows to save on hardware purchases and maintenance. The main functionalities include website execution, software application support, data analysis and storage and software development and computing tests. This kind of service is mainly used by the IT administrators within a company (Fig. 2.4).

Platform as a service (PaaS) provides access to a cloud-based environment where the back-end provider provides the underlying infrastructure for users to develop and distribute applications. In addition to storage and other computing resources, users can use a suite of predefined tools to develop and distribute their applications without wasting time in writing the source code. This type of service does not provide direct control over the infrastructure, as the providers are in charge of managing security and backups. This cloud-enabled service enables a wide range of applications to be developed, like APIs (application programming interfaces), tools for IoT deployment and business insights based on data analytics. The recipients of the cloud platform are mainly software houses or web agencies, which can collaborate in the development of apps also remotely thanks to the Internet connection.

Software as a service (SaaS) provides software and applications via third party over the Internet, to which users subscribe and can access from everywhere through the Web or the back-end vendor's API. A generic but exhaustive example is the use

Table 2.4 The strategic impacts of Industrial Internet of Things (IIoT)

Building blocks	Operational benefits	Technological strategies ("o" indicates a neutral impact)							
		Decentralization	Centralization	Externalization	Internalization	Interoperability	Ownership	Usability	Security
R	Enables the real-time location of materials and products along the supply chain		X	X		X		o	o
IP	Increases efficiency in supply chain management by continuously integrating more detailed information from the various partners in the supply chain		X	X		X		o	
IP	Increases work productivity by continuously monitoring working conditions	o		o		o		o	
IP	Increases the productivity of machinery and equipment by continuously monitoring the processing parameters	o		o		o		o	
IP	Reduces unexpected failures and the amount of preventive maintenance for machinery	o		o		o		o	

EP	Increases the accuracy of the verification of the conditions of transport and storage of the products	o		o		o		o
P	Reduces unexpected failures and the amount of preventive maintenance for products	o		o		o		o
P	Increases communication and interaction between intelligent products (through decision-making capabilities) and centralized controlling units		X	X		X		o
P	Increases the offer of personalized and automated services by leveraging IoT data	o		o		o		o
P	Reduces product repair time by signalling the need for maintenance before the product breaks	o		o		o		o

(continued)

Table 2.4 (continued)

C	Reduces the role of intermediaries in the relationship with the customer		X		X	o	o
C	Increases the simplicity of use of the product by reducing human intervention	o		o		o	o

Technological strategies ("o" indicates a neutral impact)								
Prediction	Reactivity	Augmentation	Automation	Flexibility	Efficiency	Complexification	Simplification	Value proposition
	X	o		o		o		• Accessibility, availability and rapidity
	X	o			X	o		• Price • Reliability • Accessibility, availability and rapidity
o		o			X	o		• Price
o		o			X	o		• Price
X			X		X	o		• Price • Reliability
X			X		X	o		• Accessibility, availability and rapidity
X			X		X	o		• Performance
o			X	o		X		• Performance • Newness • Complementary goods, services and experiences
X			X	X		X		• Complementary goods, services and experiences
X			X		X	o		• Reliability • Complementary goods, services and experiences
o		o		o		o		• Involvement and sharing • Complementary goods, services and experiences
o			X	o		o		• Complementary goods, services and experiences

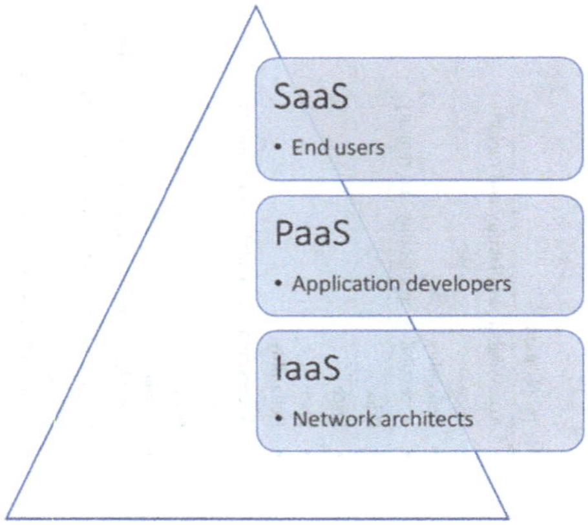

Fig. 2.4 The cloud infrastructure services

of Gmail, CRM (customer relationship management) or ERP (enterprise resource planning) software. This service is configured through the "rental" of the use of the app. Through a subscription, users can store data, analyse data or collaborate on projects.

The first advantage of using the cloud is a reduction in the investment cost needed to purchase hardware and software and to manage the local data centres. The second advantage lies in the offer of services in self-service and on demand, thus increasing the speed of accessibility to data saved in the cloud. The third advantage lies in the scalability of the services, accessible any time and from any device. The fourth advantage is represented by productivity, because on the back-end user's side, the fact of not having to manage low-level management activities allows him/her to focus on the core activities and therefore reach the market faster. Finally, the fifth advantage is that the centralized management of virtualized resources offers high security standards and protocols, which prevent cyberthreats [20, 25, 34, 38, 65, 66].

From the user's perspective, cloud computing is considered a technological paradigm as it offers an innovative approach to manage and deliver services over the Internet in order to solve, for example, data security or remodulation issues based on the use of the computational power of the cloud. Consequently, computing is being recognized more and more as a utility resource and integrated in an increasing number of complex decisions within an organization. A number of enabling technologies underlie the cloud computing paradigm, like hardware technologies, server and storage virtualization to allow different clients to access the network-enabled services; distributed computing technologies, utility and grid computing to achieve solutions autonomously; and Internet technologies—Web 2.0 and web services to enhance collaboration and information sharing on the Web [67, 68].

The paradigm of cloud computing within the **resource** management system leads to significant operational benefits, first the provision of all functionalities of the cloud servers, although drastically reducing the investment needed to implement IT services internally. This means reducing the risk of technological obsolescence.

Companies can therefore use computational tools online and on demand, accessing hardware resources almost immediately, leading to faster time-to-market. For example, for SaaS services, the required applications are run on the cloud, eliminating the need to install and run the application on the client's computer. With cloud computing, it is also possible to share resources and information through the network instead of having local servers and personal devices. Users (companies) pay the service in the form of an operational expense (pay-per-use) without incurring significant initial capital charges. The shift to the cloud will lead to a sustainable choice in terms of energy consumption. Having a cloud system allows companies to distribute resources through the value chain, automating storage backups and offering up-to-date monitoring and error reporting systems. This certainly facilitates the management of the business process, providing governance, performance, interoperability and the most appropriate security framework [5, 15, 24, 25, 58, 65, 69, 70].

With regard to **internal processes**, cloud-based collaboration tools simplify collaboration among teams even from remote locations, reducing time and costs. This happens through cloud services offering platforms that allow workers to connect from anywhere. Other cloud services allow you to monitor and control processes remotely and taking advantage of a single data source [5, 32, 65].

The cloud affects **external processes** by increasing data sharing between companies or with customers who are already using cloud-based software [5, 23].

As for the **product** building block, the cloud paradigm enables the use of IT platforms (memories, servers, networks, operating system, programming language, etc.) for the development and subsequent commercialization of software applications through the PaaS service. This eliminates the costs and complexity of purchasing and managing the underlying hardware and software levels. Other services serve to structure, share and customize remote network infrastructures (IaaS) or to use third-party applications (SaaS) [65, 71, 72].

Table 2.5 shows the operational benefits associated with cloud computing and the impact of these benefits on the building blocks constituting the business model canvas, on the technological strategies and on the value proposition.

2.6 Simulation

Simulation software is used to make a virtual copy of a physical object, in order to understand its behaviour, predict its performance and eventually compare alternatives and choose the best one. The virtual model is used to simulate a set of activities by changing its variables and understand how changes will impact on the whole mechanism. In a warehouse, the simulation process can be used to observe the productive capacity of the machines and the movement of resources needed from inventory, being able to adjust the use of them and implement just-in-time manufacturing. The creation of a simulation model usually starts from the need of fixing a problem. Once the model is built, it has to be run, and outputs have to be analysed and integrated in the manufacturing process [37].

Table 2.5 The strategic impacts of cloud computing

Building blocks	Operational benefits	Technological strategies ("o" indicates a neutral impact)							
		Decentralization	Centralization	Externalization	Internalization	Interoperability	Ownership	Usability	Security
R	Reduces the cost for initial investments and subsequent maintenance of physical infrastructures and software applications	X		X		o		o	
R	Reduces the risk of technological obsolescence of physical infrastructures and software applications	X		X		o		o	
R	Reduces the cost for the use of physical infrastructures making them infrastructure as a service (IaaS)	X		X		o		o	
R	Reduces the cost for the use of software applications making them software as a service (SaaS)	X		X		o		o	
R	Reduces energy waste in the use of physical infrastructures by exploiting economies of scale	o		o		o		o	

R	Increases the inviolability of physical infrastructures and software applications by centralizing cybersecurity procedures	o		o			X		X
R	Enables the use of physical infrastructures and software applications at all times	o		o		o		o	
R	Enables the use of physical infrastructures and software applications everywhere	X		X		o		o	
R	Enables the use of physical infrastructures and software applications from any user's device	X		X		o		o	
R	Enables the use of physical infrastructures and immediately scalable software applications	X		X		o		o	

(continued)

Table 2.5 (continued)

IP	Increases remote collaboration between teams and individuals operating from remote locations thanks to virtualization of physical infrastructures and software applications	X		X		X		o	
IP	Increases process monitoring and control		X		X		X	X	
EP	Increases data sharing between sites and organizational boundaries	X		X		X		o	
P	Enables the use of IT platforms (memories, servers, networks, operating systems, programming language, etc.) for the development and distribution of software applications (PaaS)	X		X		X			X

Technological strategies ("o" indicates a neutral impact)								
Prediction	Reactivity	Augmentation	Automation	Flexibility	Efficiency	Complexification	Simplification	Value proposition
o		o			X	o		• Price
o		o			X	o		• Performance
o		o			X	o		• Price
o		o			X	o		• Price
o		o			X	o		• Price
o		o		o		o		• Performance • Immaterial quality
o		o			X	o		• Accessibility, availability and rapidity
o		o		o		o		• Accessibility, availability and rapidity
o		o		o		o		• Accessibility, availability and rapidity
o		o		o		o		• Accessibility, availability and rapidity
o		o		o		o		• Accessibility, availability and rapidity • Newness • Involvement and sharing
	X		X		X		X	• Reliability
o		o		o		o		• Accessibility, availability and rapidity • Involvement and sharing • Complementary goods, services and experiences
o		o		o		o		• Newness • Complementary goods, services and experiences

Nowadays, simulation software packages are embedded with a user-friendly interface; this is because manufacturing processes and products are evolving rapidly and keeping up with emerging needs requires to have a flexible system. Moreover, the increasing number of latent variables that affect the success or failure of a decision in the market makes it even riskier to take action. For this reason, simulation can be viewed as a long-term problem-solving methodology, with which it is possible to evaluate results of different scenarios and choose the optimal one, without incurring in risks. Once its functionality is assessed, it is easy to understand how useful a simulation tool can be when evaluating the manufacturability of a new product or assisting in the development of a new product or process or even planning the capacity of a machine. Finally, additional outcomes of a simulation process include some indexes that may be helpful when comparing performance to a benchmark, in order to track the performance of the system. Simulation techniques can be very different as they vary depending on what they are planned for. The three most popular simulation tools are the system dynamics (SD), the discrete-event simulation (DES) and the agent-based model (ABM).

System dynamics (SD) is a simulation model for framing, understanding and identifying the dynamic non-linear behaviour of complex systems. The principle is the recognition that the underlying structure of the system is as important as the determination of its behaviour. This methodology can be deployed when considering long-term horizon of the business. System dynamic modelling is used to structure real-world phenomena, in which continuous changes occur and, for this reason, the model updates every variable in small time lapses, giving positive or negative feedbacks and continuously adjusting it. These systems ideally form flow diagrams, and their inherent structure tells a lot about the behaviour of the variables. As it is a deterministic top-down approach, the main advantage of this model is its ability to predict a system's behaviour just by looking at its structure. It can be deployed in capacity planning, activity forecasting and quality management, given its macroscopic perspective.

The discrete-event simulation (DES) structures a system in a series of discrete events in determined periods of time. For this reason, it can be viewed as a step function, as opposed to a continuous function in which time is not split up. One of the main differences between DES and SD is that the first one is primarily used in discrete events, while SD involves systems characterized by continuous changes. In general, the DES system implies a queue, like the simulation of the warehouse operations. It goes very much in detail, analysing every time sequence; while SD is deterministic, this model is rather stochastic, requiring the processing of different results before identifying a certain behaviour. It can be deployed to balance the assembly lines, to schedule a system and to optimally allocate resources or for capacity planning.

The third most used simulation tool is the agent-based model (ABM), which employs a series of autonomous and independent agents or objects which interact with each other in order to achieve an objective and explain their behaviour. The most relevant difference between DES and ABM is that the last one has a bottom-up perspective, where the same objects under study are independent and active entities with a given behaviour, while DES presents a top-down process-oriented approach. As an example, the ABM system has been used to study the migratory behaviour of birds. Indeed, different from DES, the agent-based model does not require the concept of queues. They are essentially decentralized, and together they form a pattern of behaviour. It can be deployed in transportation, production planning and supply chain supervision [34, 73–75].

One advantage in implementing the simulation paradigm consists in increasing the forecasting capacity of economic, environmental and social performance of products and processes at different levels of management according to the different stages of the product's life cycle. A second advantage of virtual simulation is that it does not require physical employment of resources, rather it diminishes waste, time and effort by providing a secure and faster method to optimize manufacturing processes.

Simulation is considered a technological paradigm as it offers an innovative approach to connect traditional computer simulation tools with various areas of an organization, ultimately influencing decisions. Traditional stand-alone models, which were comprehensible only to experts, will be replaced by models embedded with user-friendly interfaces, which open new possibilities to modify the model by every actor of the value creation process. With the advent of the digital twin, the use of the simulation paradigm extends to all the stages of the product's life cycle, requiring more dynamic and adaptable models in order to provide seamless assistance and operational data connectivity [76].

In relation to the **supplier** building block, it will be possible to simulate the different costs and margins of alternative suppliers, including the simulation of the environmental costs associated with a supplier's failure, thus reducing procurement costs [77].

Applying the simulation paradigm in **resource** management will increase the availability of information regarding the shelf life of components [77, 78].

Within **internal processes**, simulation applied in the construction of complex building allows the discovery of new modelling techniques, including modularization, prefabrication and three-dimensional printing of individual components, thus speeding up design iterations and improving visualization. Advanced production processes minimize the time required for the design until the product is delivered. The use of simulation software allows reducing the risks associated with a project through the prior detection of errors, testing the impact of alternative choices for their resolution [19, 24, 29, 36, 37, 40, 53, 77, 78].

In **external processes**, the simulation software makes it possible to virtually simulate the transport of products, tracing the best travel itinerary for delivery to the customer and following the path along the road to ensure its success [13, 52].

As for the **product** building block, various benefits can be obtained through the use of advanced simulation systems. For example, digital models can preview the production of products. This improves the management, monitoring and reconfiguration of the product itself [19, 24, 29, 36, 37, 77, 79].

Table 2.6 shows the operational benefits associated with simulation and the impact of these benefits on the building blocks constituting the business model canvas, on the technological strategies and on the value proposition.

2.7 Big Data Analytics

Big data analytics involves the collection of data and its analysis and the extraction of useful information to provide insights for the business in order to make the right business decisions. It is a process that leads to deciding on a course of action based on data, transforming intangible resources into tangible value. As big data and technologies have become more accessible, there is a growing opportunity for companies to revolutionize the decision-making process, making data-driven decisions more objective. Indeed, the process itself can become more accurate and precise when decisions are based on data about the past to predict what will happen in the future. This is why analytics can drive profound changes in business processes and even capture new market opportunities. In order for this to happen, it is fundamental to incorporate new ways of managing data, going from subjective decisions based on opinions or intuition to objective decisions based on data and insights.

Data-driven decision processes are structured mainly into three stages: observation and data collection (facts, images, sounds, etc.) in terms of volume, speed and variety, processing (formatting, filtering, synthesis) and contextualization of data collected to generate information and application of information to generate knowledge (ideas, rules, procedures) that guides decisions and actions. The process begins with identifying and collecting the data coming from internal and external sources; data are then elaborated, interpreted and given significance in order to then be shared with the decision-makers across various communication channels in the form of insights; finally, valuable information is used to assess gaps in the value creation process. The data-driven decision process uses data as an input and information insights as an output, which will contribute to extending the knowledge of the company.

A proper data infrastructure lays the foundation for the development of four main analytical capabilities: descriptive, diagnostic, predictive and prescriptive. Descriptive analysis explores past events to identify historical dynamics and performances, presenting a graphic representation of a series of data; diagnostic analysis identifies the reasons for certain trends or events to identify activities that did not lead to the desired results and that need to be reviewed; predictive analysis predicts the future behaviour of users or the performance of the organization by building probabilistic scenarios based on past patterns; and finally, prescriptive analysis provides solutions or recommendations on actions to be taken in order to better manage possible future scenarios, by quantifying the effect they could have on the system. These four methods focus, respectively, on the analysis of what happened, why it happened, what could happen and what could be done. The combined implementation of these analytical capabilities will give the company the opportunity to address more

Table 2.6 The strategic impacts of simulation

Building blocks	Operational benefits	Technological strategies ("o" indicates a neutral impact)							
		Decentralization	Centralization	Externalization	Internalization	Interoperability	Ownership	Usability	Security
S	Reduces the costs (economic and environmental) of procurement by foreseeing the costs in the choice of alternative suppliers	X		X		o		o	
R	Increases the information on the residual time of use or storage of component parts also to identify those that can be reused	o		o		o		o	
IP	Increases the possibility of collaboration at the supply chain level by simulating the overall process, forecasting economic, social and environmental performance	X		X		o		o	
IP	Increases the iterations of designing new products and processes	o		o		o		o	

(continued)

Table 2.6 (continued)

IP	Reduces the product development lead time	o	o	o	o
IP	Reduces the risks of incorrect design reflecting the requests of users in the realization of the product	o	o	o	o
IP	Reduces the time and cost of reconfiguring machinery and systems by continuously adapting them to the dynamic production context	o	o	o	o
IP	Increases control by monitoring the correctness and efficiency of the production process	o	o	o	o
IP	Increases the efficiency in the use of internal handling equipment by simulating the different assembly phases	o	o	o	o
IP	Increases the efficiency of internal logistic processes by simulating inventory levels	o	o	o	o

EP	Increases the efficiency and effectiveness of external logistic processes by simulating the transport and delivery phase	o	X		o	o	
P	Reduces the risks of incorrect design by preventing potential errors through the simulation of product behaviour	o	o		o	X	
P	Increases the performance of the product by determining the optimal operating conditions of operation	o	o		o	X	
P	Enables the design of advanced environmental management services	o	o		o	o	
P	Increases control by monitoring the correctness and efficiency of the product during its entire life cycle, anticipating its real performance	o	o		o	o	

(continued)

Table 2.6 (continued)

Technological strategies ("o" indicates a neutral impact)								
Prediction	Reactivity	Augmentation	Automation	Flexibility	Efficiency	Complexification	Simplification	Value proposition
X		o			X	o		• Price
X			X		X	o		• Price • Accessibility, availability and rapidity
X			X		X	X		• Reliability • Accessibility, availability and rapidity
	X	o		X		o		• Newness
o		o		X		o		• Newness
	X	o		o		o		• Reliability • Performance
	X		X		X	X		• Price • Accessibility, availability and rapidity
X			X		X	o		• Price • Accessibility, availability and rapidity
X			X		X	o		• Price • Accessibility, availability and rapidity
X			X		X	o		• Price • Accessibility, availability and rapidity
X			X		X	o		• Price • Accessibility, availability and rapidity
X			X	X		o		• Reliability • Performance
X			X	o		o		• Performance
o			X	o		o		• Performance • Immaterial quality • Complementary goods, services and experiences
X			X	o		o		• Performance

complex business problems and to provide solutions in a more effective way. Data-driven decision provides actionable plans to develop a faster response to company's challenges, positively influencing the overall performance [80].

There are many benefits that can be obtained from an accurate elaboration of data, regarding both production efficiency and customer strategy. Using data-driven decisions in the production process can support a more robust product design and planning, by simulating customer future reactions and thus programming the optimal amount of stock assortment based on statistics. By analysing real-time data concerning the production processes, it is possible to programme a more stable production line, avoiding maintenance costs and downtime. Furthermore, it is possible to predict the peak demand and thus guarantee the procurement of resources from the warehouse before production begins. The use of intelligent machines that send data and find optimal solutions helps reduce material and energy waste. Big data analytics also favour an optimal customer strategy by offering customer segmentation insights to better target and define the product's features. The types of data that can be extracted are about demographics, product usage and distribution channel preferences. A further application concerns the interaction with the final consumer. Data from smart products allows the company to be more informed about consumers' preferences, thus being able to propose solutions based on their purchase propensities. In order to guarantee this level of services, a rapid response from the system is necessary, and it is for this reason that analytics are performed in real time. In addition to this, another benefit that comes with the elaboration of data is the possibility to embed smart devices with an edge computing system to preprocess data at the edge of the network and then send them directly to the central cloud computing unit. The advantage of using decentralized data-driven decision can be understood by considering three factors. First, the sensitive information that is captured by edge analytics is sent to the cloud only if certain privacy conditions are satisfied, in order to preserve safety. Second, the smart devices that come with analytics are able to lighten the work on the cloud, thus reducing data latency. Third, having a decentralized and independent processing system with respect to the cloud can prevent system connection issues, which can occur especially when some devices are located in remote places [9, 19, 27, 28, 34, 45, 81–83, 80–86].

2.7.1 Descriptive Analytics

Descriptive analytics applied to **resources** will bring advantages in the implementation of autonomous robots and machine learning, thanks to the possibility of doing advanced analysis on historical data and statistical techniques to increase knowledge. Based on big data analysis, it is possible to establish an accurate knowledge of the production process and inventory; remotely track product usage, performance and positioning; and finally optimize the distribution of products [5, 40, 46, 87].

The integration of new data sources, the collection, the evaluation and the ability to perform big data analytics can help to plan **internal processes** to support real-time decision-making. Analytics allow organizations to identify customer preferences, generating ad hoc insights to simplify users' unnecessary tasks and thereby increase the customer experience. Moreover, the historical analysis of data allows to quantify the performance of machines, products and systems. This allows you to make decisions faster [5, 24, 39, 47, 85, 88–90].

With regard to **external processes**, it is possible to exploit descriptive analytics to integrate external market data to improve sales price decisions and find potential common points and opportunities to negotiate better prices [40, 85].

Products can benefit from the descriptive capabilities of data to be constantly tracked along the supply and production chain for the purpose of realizing more precise demand planning [40].

Table 2.7 shows the operational benefits associated with big data analytics (descriptive analytics) and the impact of these benefits on the building blocks constituting the business model canvas, on the technological strategies and on the value proposition.

2.7.2 Diagnostic Analytics

Diagnostic analysis predominantly impacts **internal processes**. Through the correct interpretation of big data, it is possible to diagnose past errors and downtime to plan an error-free process, realizing at the same time the desired performance levels. For example, by correlating data during product testing, it is possible to identify models that help identify faulty products within the early stages of the production process, thus improving production quality. In fact, the prognostic systems integrated in the machines include diagnostic capabilities to assess the probability of failures. Advanced analyses can promptly verify defects, failures and deficiencies by performing trend analyses to assess a more efficient use of resources and perform predictive maintenance [5, 24, 40, 41, 85, 88].

Table 2.8 shows the operational benefits associated with big data analytics (diagnostic analytics) and the impact of these benefits on the building blocks constituting the business model canvas, on the technological strategies and on the value proposition.

2.7.3 Predictive Analytics

Based on the analysis of big data, it is possible to use predictive analytics at the level of **resources** to establish an accurate knowledge of the production process and predict the use of raw materials, minimizing production as well as product redundancy [87].

From the point of view of **internal processes**, predictive analytics allow for predictive maintenance of production plants. Data sent through the sensors are used by the system to learn the regular behaviour of machines, comparing the actual behaviour with the expected behaviour to identify any anomalies in the monitored

Table 2.7 The strategic impacts of big data analytics (descriptive analytics)

Building blocks	Operational benefits	Technological strategies ("o" indicates a neutral impact)							
		Decentralization	Centralization	Externalization	Internalization	Interoperability	Ownership	Usability	Security
R	Increases quality of the tracking of goods along the supply chain for demand planning	o		o		o		o	
IP	Increases quality of insight generation that helps solve customer issues and increases customer experience	o		o		o		o	
IP	Increases quality in quantifying the performance of machines, products and systems to make decisions more quickly	o		o		o		o	
EP	Increases quality in the search for purchasing commonalities to negotiate better pricing	o		o		o		o	
P	Increases quality of the tracking of product usage, performance and positioning	o		o		o		o	

(continued)

Table 2.7 (continued)

Technological strategies ("o" indicates a neutral impact)								Value proposition
Prediction	Reactivity	Augmentation	Automation	Flexibility	Efficiency	Complexification	Simplification	
X		o		o		o		• Accessibility, availability and rapidity
X		o		o		o		• Product range and customization • Complementary goods, services and experiences
X		o			X	o		• Reliability • Accessibility, availability and rapidity
X		o			X	o		• Price
X		o		o		o		• Performance • Complementary goods, services and experiences

Table 2.8 The strategic impacts of big data analytics (diagnostic analytics)

Building blocks	Operational benefits	Technological strategies ("o" indicates a neutral impact)							
		Decentralization	Centralization	Externalization	Internalization	Interoperability	Ownership	Usability	Security
IP	Increases rapidity of tracking defects, failures and deficiencies	o		o		o		o	
IP	Reduces costs by guaranteeing a production process without errors	o		o		o		o	
IP	Reduces costs by estimating the residual useful life of the machines	o		o		o		o	
IP	Increases quality of risk management systems	o		o		o		o	

(continued)

Table 2.8 (continued)

Technological strategies ("o" indicates a neutral impact)								
Prediction	Reactivity	Augmentation	Automation	Flexibility	Efficiency	Complexification	Simplification	Value proposition
	X	o			X	o		• Price • Reliability • Performance
	X	o			X	o		• Reliability
	X	o			X	o		• Reliability
	X	o			X	o		• Reliability • Material quality

equipment. This can also lead to proactive detection and quality control to detect defects throughout the life cycle of plants and products.

Predictive analytics, integrated within a company's **external processes**, allows to anticipate market demand by integrating and analysing feedback information, social networks, etc. in real time, therefore increasing the accuracy of market profiling and segmentation. Customers' intelligence is also used to develop smart strategies based on customers' data to anticipate their need for spare parts [24, 39–41].

Predictive analytics can significantly influence **product** development, using in-depth information on current products to increase the differentiation of products or services [23, 27, 88, 89].

From the **customer**'s point of view, the information collected can be used by predictive analytics systems to understand customer needs and anticipate future preferences to shape future products [24, 40, 88].

Table 2.9 shows the operational benefits associated with big data analytics (predictive analytics) and the impact of these benefits on the building blocks constituting the business model canvas, on the technological strategies and on the value proposition.

2.7.4 Prescriptive Analytics

Prescriptive analytics offers numerous benefits to **internal processes**, including anticipating design flaws and correcting them, designing products and simulating their use based on the costs of procurement and evaluating product design options based on producibility to suggest the best alternative. The huge amounts of data available make it possible to continuously analyse and optimize production, allowing for a better use of energy consumption. All integrated systems will follow the decisions validated by prescriptive analytics to implement decisions based on the signals received [5, 20, 24, 34, 36, 40, 55, 88, 91].

In relation to **external processes**, prescriptive analytics contributes to reducing delivery lead time, proactively planning delivery operations and distributing products directly to customers. The system is also able to modify routes for distribution vehicles in real time based on unexpected events, giving customers the possibility to monitor the status of the delivery based on their exact position [24, 39, 40, 85].

Table 2.10 shows the operational benefits associated with big data analytics (prescriptive analytics) and the impact of these benefits on the building blocks constituting the business model canvas, on the technological strategies and on the value proposition.

Table 2.9 The strategic impacts of big data analytics (predictive analytics)

Building blocks	Operational benefits	Technological strategies ("o" indicates a neutral impact)							
		Decentralization	Centralization	Externalization	Internalization	Interoperability	Ownership	Usability	Security
R	Increases the accuracy of determining the raw materials needed before production to guarantee a stable product quality level	o		o		o		o	
IP	Increases the speed of adjusting the regular behaviour of machines with their expected behaviour	o		o		o		o	
IP	Increases the speed of activating quality control for the maintenance of machinery and products	o		o		o		o	
EP	Increases the ability of developing strategies based on customer data to anticipate their need for spare parts	o		o		o		o	

P	Increases the ability of differentiating and developing innovative products and services	o	o	o	o
C	Increases the ability of providing better services thanks to the identification of future customer behaviour	o	o	o	o
C	Increases the ability of improving the user's service experience	o	o	o	o
C	Increases the capability of uncovering patterns, tracking movements and analysing customer preferences to make more accurate market profiling and segmentation	o	o	o	o

(continued)

Table 2.9 (continued)

Technological strategies ("o" indicates a neutral impact)								
Prediction	Reactivity	Augmentation	Automation	Flexibility	Efficiency	Complexification	Simplification	Value proposition
X			X		X	o		• Price • Accessibility, availability and rapidity
X			X		X	o		• Reliability • Accessibility, availability and rapidity
X			X		X	o		• Reliability • Accessibility, availability and rapidity
X			X		X	o		• Accessibility, availability and rapidity • Complementary goods, services and experiences
X		o		o		o		• Newness • Product range and customization
X		o		o		o		• Complementary goods, services and experiences
X		o		o		o		• Involvement and sharing • Complementary goods, services and experiences
X		o		o		o		• Product range and customization • Complementary goods, services and experiences

Table 2.10 The strategic impacts of big data analytics (prescriptive analytics)

Building blocks	Operational benefits	Technological strategies ("o" indicates a neutral impact)							
		Decentralization	Centralization	Externalization	Internalization	Interoperability	Ownership	Usability	Security
IP	Increases rapidity in response to market demand	o		o		o		o	
IP	Increases real-time planning of energy consumption savings during breaks, reducing machine consumption	o		o		o		o	
EP	Increases the ability to modify routes of vehicles in real time based on unforeseen events	o		o		o		o	
P	Increases the capability to anticipate design flaws and correct them	o		o		o		o	

Technological strategies ("o" indicates a neutral impact)								
Prediction	Reactivity	Augmentation	Automation	Flexibility	Efficiency	Complexification	Simplification	Value proposition
	X		X	X		o		• Accessibility, availability and rapidity
	X		X		X	o		• Price
	X		X	X		o		• Accessibility, availability and rapidity
	X		X	o		o		• Performance

2.8 Cybersecurity

As products and production processes become more connected and distributed, the increasing flow of information exposes the company to the risk of cyberattacks. For this reason, confidentiality, authenticity, integrity and privacy must be ensured. Cybersecurity guarantees protection of potential damage to hardware and software components and from data breaches. Therefore, building a trusted data management infrastructure is fundamental to enhancing security and trustworthiness of computing platforms. Protecting data and making the security system resilient must be the primary objective when pursuing cybersecurity-related actions. These actions must be then reported to the institutions, which issue security regulations, and to the final consumers, which must be aware of the risks and must ensure that their privacy is protected.

An effective data management must ensure the security of data, which means that data cannot be hacked. A trusted data management is divided into two main strands: on the one hand, the nature of data is made unalterable thanks to cryptography; this means that data is certified and authenticated to preserve its value and quality; and on the other hand, the transmission of data is protected by making it non-interceptable and therefore confidential; this prevents cyberattacks when relevant information is communicated from one side of the company to the other. The fundamental activities involved in cybersecurity are the identification of the current vulnerabilities, the construction of an efficient infrastructure to protect data, the detection of cybersecurity threats, the real-time response and the recovery with minimum side effects. Cybersecurity can be applied at the production level, where data are shared in order to coordinate production activities, ensuring confidentiality of information; at the cloud level, where data are stored and made accessible even to third-party suppliers, protecting relevant information from unauthorized access; and finally, at the user level, where information about product usage are sent through smart products, providing a user authentication system to validate his/her identity. A company must therefore align all of its divisions towards the concept of cybersecurity, not in a fragmented way but rather with a cohesive point of view.

Significant benefits in implementing cybersecurity actions are achieved by forming collaborative partnership with third parties. Sharing data management best practices in a community can help on the one hand to raise awareness about the issue of security and on the other hand to set new standards for a secure and connected cyber-physical production system. Even though security, privacy and data sharing can be seen as potential threats towards the adoption of connected systems, it is actually an opportunity for the company to communicate transparency and attract the customer's attention, by providing safe and reliable forms of communication to and from end users. In particular, data transferred to the cloud computing storage becomes relevant in the field of cybersecurity, as protective actions must be taken both at the edge of the network and at the central level. The benefits of using the cloud are valid also for cybersecurity technologies, as it provides a more rapid response to cyberattacks and integration at all levels [4, 7, 34, 35, 82, 92, 93].

Cybersecurity is considered a digital paradigm as it offers an innovative approach to protecting data, applications and infrastructures of a data-centric company as an integrated solution for businesses and not as an overlay. This solution has the strategic objective of shortening the time between a potential threat detection and an appropriate response to it. Specifically, a cybersecurity framework is composed by five steps: identify, protect, detect, respond and recover. Some examples of technologies related to the digital paradigm of cybersecurity include backups, data encryption, storage protection, authentication and access, public key infrastructure, etc. [94].

Building a trusted data management system helps companies secure their **resources**, for example, from theft or deletion of data from servers. This helps to better manage end-to-end IT security to protect operations and proprietary data. A data security infrastructure allows you to control who has access to data at the time of the transfer, in order to avoid the release of copyrighted material on the Internet. Making security an integral part of all systems and processes represents the goal and the main advantage of cybersecurity, which also makes it possible to speed up response time in the event of information attacks [18, 24, 27, 61, 63].

The benefits in terms of **internal processes** reside in the continuous monitoring of working conditions and process parameters to detect accidents more effectively. It is essential for a cybersecurity infrastructure to keep track of internal activities and processes so that you can easily retrace and identify any unwanted events. In addition, stronger situational awareness is developed among employees which helps to better understand harmful behaviours that lead to potential cyber risks [23, 24, 63].

For what concerns **products**, computer security systems are able to identify in advance any product malfunctions caused by cyberattacks. In fact, even the products contain distinctive and inimitable information that needs to be protected from unauthorized external access [15, 17, 23, 24, 63].

Table 2.11 shows the operational benefits associated with cybersecurity and the impact of these benefits on the building blocks constituting the business model canvas, on the technological strategies and on the value proposition.

2.9 Structural Horizontal and Vertical System Integration

Structural horizontal and vertical system integration is the last paradigm marking the advent of the Fourth Industrial Revolution. Industry 4.0 demands for greater integration of data and systems along the entire value chain so that all departments and business functions become part of a single integrated system. The company will become more cohesive and truly automated. Structural horizontal integration refers to the coordinated management of the production chain; vertical integration consists of the coordinated management of the product's life cycle.

Horizontal and vertical integrations converge at a certain point; the vertical integration of hierarchical subsystems leads to the creation of an intelligent factory, which in turn supports structural horizontal integration through valuable networks.

Table 2.11 The strategic impacts of cybersecurity

Building blocks	Operational benefits	Technological strategies ("o" indicates a neutral impact)							
		Decentralization	Centralization	Externalization	Internalization	Interoperability	Ownership	Usability	Security
R	Reduces the risk of proprietary data theft intended to exploit data for competitive purposes	o		o		o		X	
R	Reduces the risk of theft of proprietary data in order to make them public	o		o		o		X	
IP	Reduces the risk of data corruption that could compromise workplace safety	o		o		o		X	
P	Reduces the value of sensitive data if compromised by increasing cloud storage and protection	o		o		o		X	
P	Reduces the risk of data corruption that could create malfunction in products	o		o		o		X	

Technological strategies ("o" indicates a neutral impact)								
Prediction	Reactivity	Augmentation	Automation	Flexibility	Efficiency	Complexification	Simplification	Value proposition
o		o		o		o		• Immaterial quality
o		o		o		o		• Immaterial quality
o		o		o		o		• Reliability
o		o		o		o		• Performance • Immaterial quality
o		o		o		o		• Material quality • Performance

Thanks to structural horizontal system integration, the company will be able to exchange information with decentralized production sites, fostering the creation of global value chain networks, an integrated ecosystem of inter-company collaborations. Instead, vertical system integration will enable access to automated production equipment, achieving the potential of a smart production system by using sensors to build a reconfigurable manufacturing system. The combination of vertical and horizontal integrations can increase the productivity of companies, with increased resource utilization in terms of efficiency, better product-customer match and, finally, leading to the generation of new business models. This will lead to a transformation process that will impact not only the economic sphere but also the social sphere by changing the role of the firm with respect to its surrounding environment.

In the context of Industry 4.0, the advantage of introducing structural horizontal system integration is represented by the generation of value creation networks which involve the integration of different agents such as business partners and customers. It integrates IT systems into the different stages of the manufacturing process, within a company and among several companies. It is basically identified in the collaboration between enterprises, exchanging resources and real-time information. The advantage of utilizing vertical integration involves having intelligent production systems such as factories and intelligent products, smart logistic networking, production, marketing and services. It is the integration between two hierarchical levels: production and management [1, 4, 9, 34–36, 87, 95–98].

Suppliers will undergo a major change due to vertical and horizontal integration, specifically with regard to supply chain management. Indeed, conventional supplier-buyer relations will be transformed and move towards a joint design of products and services. This collaboration is mutually beneficial for better coordination and an overall reduction in operating costs [14, 39, 64, 99–101].

The paradigm of horizontal and vertical integration applied to **resources** increases the capacity for collaboration between the various parts of the supply chain, leading to effective coordination between inventory levels and delivery times within the value chain. However, an integrated digital supply chain not only has optimized physical flows but also information flows to regulate data transmission. However, even physical objects, in addition to human resources, can communicate with each other. The processes and data sharing will become common in factories that decide to integrate their resources end to end, making them available to all partners in the value chain. Through the concept of horizontal integration, collaborative networks between companies combine resources, divide risks and adapt quickly to market changes, seizing new opportunities [27, 39, 95].

In relation to **internal processes**, vertical integration allows production planning in real time, increasing efficiency in machine use and faster production times. By vertically integrating different organizational levels with information coming from different departments, it is possible to provide managers with up-to-date information on production that serves to make the decision-making process more objective. All data on operational processes will then be optimized in an integrated network and shared between a completely digitalized partner ecosystem integrated with virtualized processes, favouring decentralized autonomy. Real-time access to an extensive series of operational information makes it possible to track the positioning of physical assets to ensure correct handling in the production processes. Intelligent machines can be dynamically reconfigured to adapt to new decisions that are made through information that is constantly collected and processed [14, 19, 27, 70, 87, 95, 102].

The integration paradigm impacts **external processes**, providing greater agility on delivery and a reduction in time-to-market. This is possible because of a more efficient division of labour leading to increased flexibility and cooperation. Vertical integration allows to distinguish the different preferences of the individual customer and customize the sales experience through contact points. Collaboration is an advantage for company and customers, who communicate through digital channels and intelligent products [32, 60, 95, 99].

With regard to **products**, integration allows for better integration and interaction between intelligent objects, performing complex cooperation tasks, but also among intelligent objects and companies, which can control quickly and effectively the performance of products [19, 102].

Structural horizontal and vertical system integration will also benefit **customers** in terms of customer support services, transforming relationships into a virtual collaboration, increasing the quality of the purchasing experience and improving the processing times of information from users [99–101].

Table 2.12 shows the operational benefits associated with structural horizontal and vertical system integration and the impact of these benefits on the building blocks constituting the business model canvas, on the technological strategies and on the value proposition.

The ongoing digital acceleration affecting all industries has led to the emergence of the aforementioned technologies. As technologies become digitized, they enter a period of deceptive growth, meaning that as technologies become information-enabled and information fuelled, their price and performance double yearly. Once the doubling pattern begins, it does not cease. The rapid increase characterizing digital technologies describes technological disruption. Once information-enabled technologies become disruptive, they dematerialize such as apps onto smartphones. Once dematerialization has occurred, the product or service demonetizes. The final stage of the process involves democratization, as access to technological advancements becomes accessible to a wide audience [62, 103].

2.10 Analysis of the Impact of Operational Benefits

In the previous paragraphs, we have identified a total of 133 operational benefits deriving from the adoption of nine technological paradigms. An analysis of their impact on the value proposition, on the building blocks constituting the business model canvas and on the technological strategies, is reported below.

In relation to the value proposition (see Fig. 2.5), the most relevant impact of operational benefits concerns the domain of operational excellence:

Table 2.12 The strategic impacts of structural horizontal and vertical system integration

Building blocks	Operational benefits	Technological strategies ("o" indicates a neutral impact)							
		Decentralization	Centralization	Externalization	Internalization	Interoperability	Ownership	Usability	Security
S	Increases the traceability of data horizontally along the supply chain thanks to signalling systems	X		X		X		o	
S	Increases information sharing with suppliers	X		X		X		o	
S	Increases the flexible and dynamic interaction between partners in the production process	X		X		X		o	
R	Increases collaboration flexibility by optimizing physical flows (inventory levels) and digital flows (information)	X		X		X		o	
R	Increases efficiency in the use of resources thanks to sharing between collaborative networks	X		X		X		o	

IP	Reduces production times by improving the use of machinery thanks to real-time planning	o	o	o	o
IP	Increases decision-making support based on up-to-date production information and incoming orders	o	o	o	o
IP	Increases real-time access to an extended set of operational information in a digitized partner ecosystem	X	X	X	o
IP	Increases the capacity for self-organization and flexible reconfiguration of production lines for customized products	o	o	o	o
IP	Increases the speed of adaptation to different types of events (e.g. production line interruption)	o	o	o	o

(continued)

Table 2.12 (continued)

EP	Reduces delivery times by coordinating and adapting production and distribution processes	o	o	o	o
EP	Reduces time-to-market	o	o	o	o
EP	Increases control over digital channels and smart products to improve the customer experience	X	X	X	o
P	Increases interaction between products to perform complex cooperation tasks	o	o	X	X
P	Increases the interaction between the company and its products for a more effective control and a constant flow of information	o	o	o	o
C	Increases customer relations by facilitating joint product design	o	o	o	X

Technological strategies ("o" indicates a neutral impact)								
Prediction	Reactivity	Augmentation	Automation	Flexibility	Efficiency	Complexification	Simplification	Value proposition
	X		X	X		o		• Accessibility, availability and rapidity
	X		X	X		o		• Accessibility, availability and rapidity • Material quality
	X		X	X		o		• Accessibility, availability and rapidity
	X		X	X		o		• Price • Accessibility, availability and rapidity
o			X		X	o		• Price
	X		X		X	o		• Price • Accessibility, availability and rapidity
	X		X	o		o		• Accessibility, availability and rapidity
	X		X	o		o		• Accessibility, availability and rapidity
o		o		X		o		• Price • Accessibility, availability and rapidity
o		o		X		o		• Accessibility, availability and rapidity
o		o			X	o		• Accessibility, availability and rapidity
o		o			X	o		• Newness
o		o		o		X		• Involvement and sharing • Complementary goods, services and experiences
o			X	X		X		• Performance • Newness • Complementary goods, services and experiences
	X		X	o		o		• Performance
o		o		o		o		• Involvement and sharing

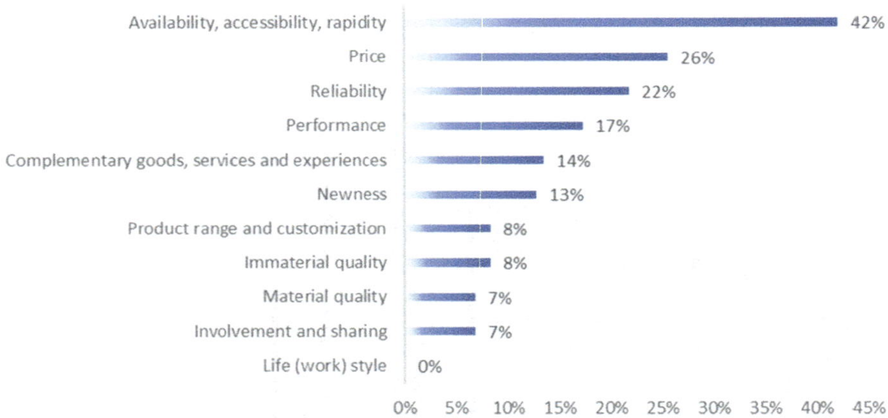

Fig. 2.5 The impact of operational benefits on the value proposition

- Forty-two percent of the 133 operational benefits listed above have an impact on accessibility, availability and rapidity.
- Twenty-six percent have an impact on price.
- Twenty-two percent have an impact on reliability.

The other major impact concerns product leadership:

- Seventeen percent of the 133 operational benefits listed above have an impact on performance.
- Thirteen percent have an impact on newness.
- Eight percent have an impact on immaterial quality.
- Seven percent have an impact on material quality.

The less frequent impacts concern both customer intimacy and ecosystem leadership.

As far as the building blocks constituting the business model canvas are concerned, there is a significant prevalence of the impact on internal processes, as shown in Fig. 2.6. Consistent with the results obtained regarding the impacts on the value proposition, the operational benefits of Industry 4.0 mainly benefit internal processes.

Finally, the other two areas of the business model canvas that are affected are products and resources, although less compared to the remaining building blocks. These results show that the value chain area, that is, resources—(internal and external) processes—products, is most impacted by Industry 4.0.

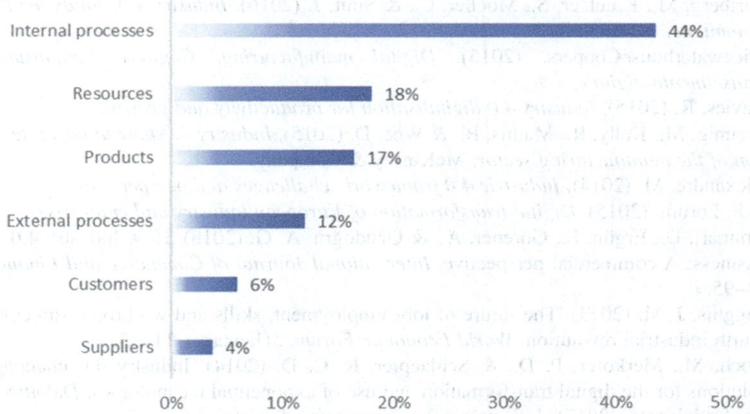

Fig. 2.6 The impact of operational benefits on the remaining building blocks

References

1. Bartodziej, C. J. (2017). *The concept Industry 4.0.*
2. Dosi, G., & Nelson, R. R. (2016). Technological paradigms and technological trajectories. *Palgrave Encyclopedia of Strategic Management, 1982*, 1–12.
3. Martinelli, A. (2019). *Tecnologico, paradigma in "Dizionario di Economia e Finanza"*. http://www.treccani.it/enciclopedia/paradigma-tecnologico_%28Dizionario-di-Economia-e-Finanza%29/
4. Grosvenor, R., Liu, Y., & Qin, J. (2016). A categorical framework of manufacturing for industry 4.0 and beyond. *Elsevier, 52*, 173–178.
5. The Boston Consulting Group. (2015). *Industry 4.0: The future of productivity and growth in manufacturing industries*.
6. Cordeiro, G. A., Ferro, R., & Ordóñez, R. E. C. (2019). Theoretical proposal of steps for the implementation of the Industry 4.0 concept. *Brazilian Journal Operations Production Management, 16*(2), 166–179.
7. PricewaterhouseCoopers. (2015). *Digital manufacturing. Cogliere l'opportunità del rinascimento digitale.*
8. Lima, F., et al. (2019). Digital manufacturing tools in the simulation of collaborative robots: Towards Industry 4.0. *Brazilian Journal of Operations Production Management, 16*(2), 261–280.
9. Ganzarain, J., Ibarra, D., & Igartua, J. I. (2018). Business model innovation through Industry 4.0: A review. *Procedia Manufacturing, 22*, 4–10.
10. Kleer, R., Piller, F. T., & Weller, C. (2014). Business models with additive manufacturing. *Advances in Production Technology.*
11. Dong, H., El Saddik, A., & Zhang, L. (2015). From 3D sensing to printing. *ACM Transactions on Multimedia Computing Communications and Applications, 12*(2), 1–23.
12. Berger, R. (2016). Industrie 4.0. *"How it reshuffles the economic, social and industrial mode"*.
13. Gawankar, S. A., Gunasekaran, A., & Kamble, S. S. (2018). Sustainable Industry 4.0 framework: A systematic literature review identifying the current trends and future perspectives. *Process Safety and Environmental Protection, 117*, 408–425.
14. Ernst & Young. (2017). *Roadmap for Industry 4.0.*

15. Carlberg, M., Kreutzer, S., Moeller, C., & Smit, J. (2016). *Industry 4.0: Study for the ITRE Committee.*
16. PricewaterhouseCoopers. (2015). *Digital manufacturing. Cogliere l'opportunità del rinascimento digitale.*
17. Davies, R. (2015). *Industry 4.0 digitalisation for productivity and growth.*
18. Breunig, M., Kelly, R., Mathis, R., & Wee, D. (2015). *Industry 4.0-how to navigate digitization of the manufacturing sector.* McKinsey & Company.
19. Alexandre, M. (2014). *Industrie 4.0 framework, challenges and perspectives.*
20. S. P. Forum. (2015). *Digital transformation of European industry and enterprises.*
21. Ammari, D., Ergün, E., Görener, A., & Özüdoğru, A. G. (2018). How Industry 4.0 changes business: A commercial perspective. *International Journal of Commerce and Finance, 4*(1), 84–95.
22. Higgins, J. M. (2013). The future of jobs employment, skills and workforce strategy for the fourth industrial revolution. *World Economic Forum, 5*(January), 11–23.
23. Koch, M., Merkofer, P. D., & Schlaepfer, R. C. D. (2014). Industry 4.0 challenges and solutions for the digital transformation and use of exponential technologies. *Deloitte.*
24. Confartigianato. (2017). *Industria 4.0 e manifattura digitale.*
25. Bechtold, J., Bernhofer, L., Kern, A., & Lauenstein, C. (2011). Industry 4.0: The Capgemini consulting view. *Capgemini Consulting Digital Transformation of Supply Chains.*
26. Eddy, S., & Misthal, B. (2013). R*ethinking innovation in industrial manufacturing.*
27. Geissbauer, R., Schrauf, S., & Vedso, J. (2016). Industry 4.0: Building the digital enterprise. *Web.*
28. Bolle, M., Lorenz, M., Lueth, K. L., Strack, R., & Rüßmann, M. (2015). *Man and machine in Industry 4.0.* The Boston Consulting Group.
29. Bremicker, M., & Heynitz, H. v. (1990). *The factory of the future.*
30. Berman, B. (2012). 3-D printing: The new industrial revolution. *Business Horizons, 55*(2), 155–162.
31. Klincewicz, K. (2019). Robotics in the context of industry 4.0: Patenting activities in Poland and their comparison with global developments. *Problemy Zarzadzania, 2*(82), 53–95.
32. Microsoft. (2019). *2019 manufacturing trends report.*
33. Seghezzi, F. (2015). *Come cambia il lavoro nell' Industry 4.0?.*
34. Adamik, A. (2019). Creating a competitive advantage in the age of industry 4.0. *Problemy Zarzadzania, 2*(82), 13–31.
35. Buliga, O., Müller, J. M., & Voigt, K. I. (2018). Fortune favors the prepared: How SMEs approach business model innovations in Industry 4.0. *Technology Forecasting & Social Change, 132*, 2–17.
36. The Boston Consulting Group. (2016). *Digital in engineering and construction.*
37. C. F. I. (2015). *CFI, Roadmap per la ricerca e l'innovazione - Research and Innovation Roadmap.*
38. Franke, J. (2015). *A look at Industry 4. 0 – How the 4th industrial revolution could improve efficiency in the plastics industry.*
39. Berttram, P., & Schrauf, S. (2016). Industry 4.0: How digitization makes the supply chain more efficient, agile, and customer-focused. *Strategy &Technology.*
40. Cotteleer, M. J., Mahto, M., & Sniderman, B. (2016). *Industry 4.0 and manufacturing ecosystems: Exploring the world of connected enterprises.* Deloitte University Press.
41. Ingwald, A., & Kans, M. (2016). Business model development towards service management 4.0. *Procedia CIRP, 47*, 489–494.
42. Barrientos, A., Crespo, E., Martín-Barrio, A., Peña-Tapia, E., & Roldán, J. J. (2019). A training system for Industry 4.0 operators in complex assemblies based on virtual reality and process mining. *Robotics Computer Integrated Manufacturing, 59*(July 2018), 305–316.
43. Ferraguti, F., et al. (2019). Augmented reality based approach for on-line quality assessment of polished surfaces. *Robotics Computer Integrated Manufacturing, 59*(April), 158–167.

44. Gadh, R., Lu, S. C. Y., & Shpitalni, M. (1999). Virtual and augmented reality technologies for product realization. *CIRP Annals – Manufacturing Technology, 48*(2), 471–495.
45. GTAI. (2013). *Smart manufacturing for the future.*
46. Baur, C. (2015). *Manufacturing's next act.* http://www.mckinsey.com/business-functions/operations/our-insights/manufacturings-next-act
47. Horváth & Partners. (2015). *How Industry 4.0 is changing how we manage value creation.*
48. Gilchrist, A. (2016). *Industry 4.0 The Industrial Internet of Things.* Apress.
49. Milan, F. (2016). *Bretonstone e l'Industry 4.0.*
50. Blanchet, M., Rinn, T., Thaden, G. von, & Thieulloy, G. de (2014). *The new industrial revolution: how Europe will succeed.*
51. Lu, Y., & Xu, X. (2018). Resource virtualization: A core technology for developing cyber-physical production systems. *Journal Manufacturing Systems, 47*, 128–140.
52. PricewaterhouseCoopers. (2015). *The Internet of Things: what it means for US manufacturing.*
53. Brettel, M., Friederichsen, N., Keller, M., & Rosenberg, M. (2014). How virtualization, decentralization and network building change the manufacturing landscape: An Industry 4.0 perspective. *International Journal of Mechanical, Aerospace, Industrial, Mechatronic and Manufacturing Engineering, 8*(1), 37–44.
54. Díaz-Garrido, E., Luz Martín-Peña, M., & Sánchez-López, J. M. (2018). The digitalization and servitization of manufacturing: A review on digital business models. *Strategic Change, 27*(2), 91–99.
55. Partners, S., et al. (2015, January). Industrial Internet of Things: Unleashing the potential of connected products and services committed to improving the state of the world. *World Economic Forum.*
56. Arnold, C., & Kiel, D. (2016). How Industry 4.0 changes business models in different manufacturing industries. *ISPIM Innovation Conference*, no. June.
57. Paine, T. (2015). *Industrial Internet of Things and communications at the edge.*
58. Anastasiadis, T., Lampropoulos, G., & Siakas, K. (2019). Internet of Things in the context of Industry 4.0: an overview. *International Journal Enterprise Knowledge, 7*(1), 4–19.
59. Raiwani, Y. P. (2013). Internet of Things: A new paradigm. *International Journal Scientific Research Publications, 3*(4).
60. Montanus, M., & Thesis, G. (2016). *Business models for Industry 4.0.* Delft University of Technology.
61. Krapež, A., Meško, M., & Roblek, V. (2016). A complex view of industry 4.0. *Sage Open, 6*(2), 215824401665398.
62. Siagri, R. (2021). *La servitizzazione. Per un futuro senza limiti di crescita.* Edizioni Angelo Guerini e Associati.
63. Deloitte. (2016). Cyber risk in advanced manufacturing.
64. Burmeister, C., Luettgens, D., & Piller, F. T. (2015). Business model innovation for industrie 4.0: Why the 'industrial internet' mandates a new perspective on innovation.
65. Bandyopadhyay, S., Ghalsasi, A., Li, Z., Marston, S., & Zhang, J. (2011). Cloud computing – The business perspective. *Decision Support Systems, 51*(1), 176–189.
66. Cao, Q., Tao, F., Wang, X., Zhang, Z., & Zhu, X. (2019). Cloud manufacturing paradigm with ubiquitous robotic system for product customization. *Robotics and Computer-Integrated Manufacturing, 60*(February), 12–22.
67. Gaud, S., Gaur, A., & Jain, A. (2017). *Advance computing paradigm with the perspective of cloud computing-an analytical study.*
68. Salama, M., & Shawish, A. (2014). *Cloud computing: Paradigms and technologies.*
69. C. Consulting. (2016). Implications of Industry 4.0 for CIOs.
70. Berger, R. (2016). Indagine conoscitiva sul modello "Industry 4.0". *EnergiaMedia.*
71. Ardito, L., Garavelli, A. C., Panniello, U., & Petruzzelli, A. M. (2019). Towards Industry 4.0: Mapping digital technologies for supply chain management-marketing integration. *Business Process Management Journal, 25*(2), 323–346.

72. Hu, P. (2018). *A holistic framework for open low-power Internet of Things technology ecosystems*.
73. Herrmann, J. W., Lin, E., Ram, B., & Sarin, S. (2000). *Adaptable simulation models for manufacturing*. PhD Proposal, 1 (March).
74. McLean, C., & Leong, S. (2008). The expanding role of simulation in future manufacturing. *Proceeding of the 2001 Winter Simulation Conference*.
75. Maidstone, R. (2012). Discrete event simulation, system dynamics and agent based simulation: Discussion and comparison. *System*, no. March, 1–6.
76. Rodič, B. (2017). Industry 4.0 and the new simulation modelling paradigm. *Organizacija, 50*(3), 193–207.
77. Henning, W., Johannes, H., & Wolfgang, K. (2013). *Recommendations for implementing the strategic initiative Industrie 4.0*.
78. Elangeswaran, C., Sanders, A., & Wulfsberg, J. (2016). Industry 4.0 implies lean manufacturing: Research activities in industry 4.0 function as enablers for lean manufacturing. *Journal of Industrial Engineering and Management, 9*(3), 811–833.
79. Mezgár, I., & Pedone, G. (2018). Model similarity evidence and interoperability affinity in cloud-ready Industry 4.0 technologies. *Computers in Industry, 100*, 278–286.
80. PwC & D. Services. (2017). *Experience design in a technology world: A human centered approach to digital innovation PwC's digital services*.
81. Hinterhuber, A., & Liozu, S. (2012). Is it time to rethink your ICT and education strategy? *MIT Sloan Management Review, 53*(4), 68–77.
82. Broberg, O., Kadir, B. A., & Souza da Conceição, C. (2019). Current research and future perspectives on human factors and ergonomics in Industry 4.0. *Computer & Industrial Engineering, 137*, 106004.
83. Brandt, D., & Cernetic, J. (1998). *Human-centred approaches to control and information technology: European experiences organization technology*.
84. Alonso, R. S., Casado-Vara, R., Corchado, J. M., Rodríguez-González, S., & Sittón-Candanedo, I. (2019). A review of edge computing reference architectures and a new global edge proposal. *Future Generation Computer Systems., 99*(2019), 278–294.
85. Henke, N., & Kaka, N. (2018). Analytics comes of age. *McKinsey Analytics*, no. January, 100.
86. H. Services. (2018). Guide to data-driven decision making, no. March, 1–84.
87. Li, D., Wan, J., Wang, S., & Zhang, C. (2016). Implementing smart factory of Industrie 4.0: An outlook. *International Journal Distributed Sensor Networks, 2016*(1), 1–23.
88. Kaynak, O., & Yin, S. (2015). Big Data for modern industry: Challenges and trends. *Proceedings of the IEEE, 103*(2), 143–146.
89. Chiang, R. H. L., Grover, V., Liang, T. P., & Zhang, D. (2018). Creating strategic business value from big data analytics: A research framework. *Journal Management Information System, 35*(2), 388–423.
90. Heo, J. Y., Lim, C., Kim, K. H., Kim, K. J., Kim, M. J., & Maglio, P. P. (2018). From data to value: A nine-factor framework for data-based value creation in information-intensive services. *International Journal Information Management, 39*, 121–135.
91. Härting, R. C., Jozinović, P., Neumaier, P., Möhring, M., Reichstein, C., & Schmidt, R. (2015). Industry 4.0 – Potentials for creating smart products: Empirical research results. *Lecture Notes in Business Information Processing, 208*, 16–27.
92. Sveiby, K. E. (1997). The intangible assets monitor. *Journal of Human Resource Costing Accounting, 2*(1), 73–97.
93. Csik, M., Frankenberger, K., & Gassmann, O. (2016). *St. Gallen Business Model Navigator*.
94. Dickson, F., & Goodwin, P. (2019). *White paper five key technologies for enabling a cyber-resilience framework updated report*.
95. Ayala, N. F., Benitez, G. B., Dalenogare, L. S., & Frank, A. G. (2018). The expected contribution of Industry 4.0 technologies for industrial performance. *International Journal Production Economics, 204*(August), 383–394.

96. Mrugalska, B., & Wyrwicka, M. K. (2017). Towards lean production in Industry 4.0. *Procedia Engineering, 182*, 466–473.
97. Chukalov, K. (2017). Horizontal and vertical integration, as a requirement for cyber-physical systems in the context of Industry 4.0. *International Scientific Journal Industry 4.0, 157*(4), 155–157.
98. Marmolejo-Saucedo, J. A., Pérez-Lara, M., Salais-Fierro, T. E., Saucedo-Martínez, J. A., & Vasant, P. (2018). Vertical and horizontal integration systems in Industry 4.0. *Wireless Networks*, no. November.
99. Geissbauer, R., Koch, V., Kuge, S., & Schrauf, S. (2014). Industry 4.0: Opportunities and challenges of the Industrial Internet. PricewaterhouseCoopers. *Aktiengesellschaft Wirtschaftsprüfungsgesellschaft, 2*(3), 52.
100. Economico, C., & Giobbe, S. (2017). Industry 4.0: cosa ne pensano le aziende italiane.
101. Kazantsev, N., Mehandjiev, N., Pishchulov, G., Sampaio, P., & Zolkiewski J. (2017). *Formation of demand-driven collaborations between suppliers in industry 4.0 production networks.*
102. Corradi, A. et al. (2018). Smart appliances and RAMI 4.0: Management and servitization of ice cream machines. *IEEE Transactions on Industrial Informatics*.
103. Ismail, S. (2014). *Exponential organizations. Why new organizations are ten times better, faster, and cheaper than yours (and what to do about it)*. Diversion Books.

Industry 4.0 and the Emergent Business Models

3

Digital shifts are opening innovative opportunities in multiple industries. The current technological revolution is transforming industries through software technologies. As Marc Andreessen affirms, 'software is eating the world', considering that, currently, numerous software companies are poised for global dominance. Major organizations operating in numerous sectors rely on software and deliver online services. Furthermore, software programming tools as well as the Internet contributed to launching innovative global software-powered start-ups, without requiring major investments in infrastructure or employee training. Organizations in all industries must assume that a software revolution is coming. Incumbent corporations appear increasingly threatened by the emergence of innovative software offerings. In this regard, advanced software ideas are going to result in the rise of technology start-ups invading existing industries [1].

Building an agile and hybrid organization may represent a valid solution in order to survive and thrive through digital disruption. Furthermore, conventional companies are transitioning from traditional hierarchies to agile enterprises for the purpose of increasing productivity and accelerating time to market, thus seizing the benefits of the digital age and triggering innovation. The aforementioned capabilities enable organizations to navigate numerous challenges faced by traditional enterprise structures, ultimately leading to enhanced productivity and greater responsiveness to fluctuations in demand.

Digital technologies contribute to increasing customer value, thus leading to entirely new business models and service offerings. The connection between business models and technology is bijective: on the one hand, technology influences business models, on the other hand, technologies necessitate business models. Although technologies enable the emergence of innovative business models, technological success and failures are connected to the business model implemented. Therefore, the advent of technological advances enables business model innovation.

Business models elicit and foster technological innovation and cause disruption, changing the rules of the game and triggering a new wave of digital innovation. As innovative as new technologies are, technological advancements have no economic

C. Bagnoli et al., *Digital Business Models for Industry 4.0*, Future of Business and Finance, https://doi.org/10.1007/978-3-030-97284-4_3

value, unless they are properly implemented within business models. A business model which is driven by the digital paradigms characterizing Industry 4.0 will use technologies as an input in order to extract value and generate a valuable output for its customer base. The operational benefits that the technological paradigms characterizing Industry 4.0 produce impact on the critical success factors underlying the value propositions and, therefore, on the strategic dimensions. The potential impact of Industry 4.0 on the critical success factors, and on the strategic dimensions, allows in turn to design new value propositions to create new business models. Starting from a few studies investigating the strategic implications of Industry 4.0, 22 potential digital business models were recognized and grouped into four meta-business models 4.0: data-driven, platform, smart factory, and servitization [2].

- Data-driven business models are typically adopted by organizations operating within the Customer-to-Business (C2B) sector, namely markets within which innovative products are designed and developed from the data generated and collected as a result of product utilization as well as product–user interactions. Data-driven business models intend to overcome the knowledge exploration versus knowledge exploitation paradox, in order to exploit existing knowledge through the generation of financial resources necessary to explore additional (customer) knowledge, while concurrently exploring new knowledge in order to generate the intellectual resources necessary to exploit existing knowledge.
- The platform business model represents a digital business model, originating from the digital operating model and data-driven business model. Platform firms adopt a Customer-to-Customer (C2C) sales model, thus enabling value-creating interactions between external producers and consumers. Platform business models intend to overcome the producer versus consumer paradox, as consumers as well as recipients of services participate in the different stages of the production process, thus assuming the role of prosumers.
- The smart factory business model constitutes a digital business model developed from the phygital operating model. Smart factories adopt the Business-to-Business (B2B) sales model, thus maximizing manufacturing efficiency as a result of the implementation of advanced technologies. Smart factory business models intend to overcome the serial production versus singular production paradox, thus manufacturing customized goods through standardized production processes.
- The servitization business model represents the final digital business model identified as a result of the present research. Organizations implementing servitization adopt a Business-to-Consumer (B2C) sales model, thus offering an innovative paradigm facilitating interactions occurring amongst manufacturers and customers in order to strengthen existing consumer relationships and secure additional revenue. Servitization business models aim to overcome the tangible product versus intangible product paradox, defined as the ability to sell intangible qualities characterizing tangible assets and tangible qualities distinguishing intangible services.

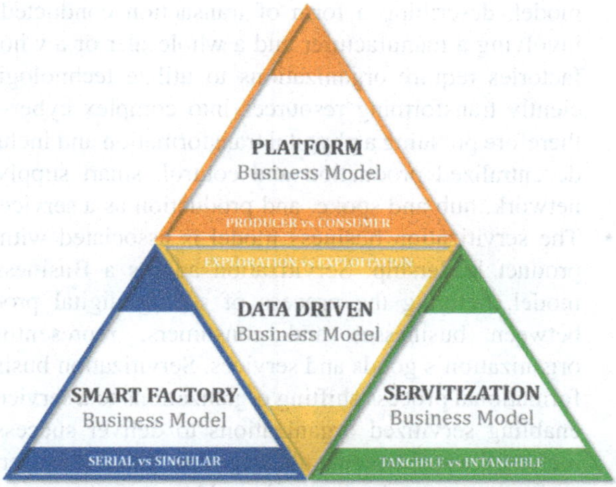

Fig. 3.1 The digital business models

The aforementioned business models are represented in Fig. 3.1.

Business models are relevant for both emerging as well as established businesses, enabling developing companies to attract investments, recruit talent and motivate management, and ensuring incumbent (non-digitized or digitalized) organizations regularly update business plans in order not to fail to anticipate trends as well as challenges.

'*In incumbent* (non-digitized or digitalized) *companies, value creation and capture are usually straightforward and closely intertwined: value is typically created and captured from the same source (the customer) through a simple pricing mechanism*' [3]. By contrast, in a fully digitized business, value creation and capture can be separated and often come from different stakeholders [4].

The Fourth Industrial Revolution has led to the emergence of advanced business models, as a result of the advancement of multiple disruptive technologies, radically transforming numerous industries. Digital business models arise as a result of the current technological revolution, innovating operational processes and user experiences as well as transforming industries.

The economic literature proposes a classification scheme identifying four distinct options, including the smart factory business model, the servitization business model, the data-driven business model, and the platform business model. Nonetheless, current research has suggested further subdivisions within each business model category, highlighting the multiple technological opportunities emerging from the on-going industrial revolution. Recent research identified multiple digital business models as well as digital operating models, characterizing Industry 4.0, categorized into four distinct classifications: smart factory business models, servitization business models, data-driven business models, and platform business models.

- The smart factory business model is associated with the strategic dimension of operational excellence. Smart factories adopt a Business-to-Business (B2B) sales

model, describing a form of transaction conducted between businesses, often involving a manufacturer and a wholesaler or a wholesaler and a retailer. Smart factories require organizations to utilize technologies for the purpose of efficiently transforming resources into complex cyber-physical goods or services, therefore pursuing a phygital transformation and include numerous subcategories: decentralized production and control, smart supply chain, agile collaboration network, hub and spoke, and production as a service.
- The servitization business model is associated with the strategic dimension of product leadership. Servitization adopts a Business-to-Consumer (B2C) sales model, defining the process of selling digital products and services directly between businesses and consumers, representing the end-users of the organization's goods and services. Servitization business models involve a transformational process shifting organizations to a service-centric business logic, thus enabling servitized organizations to deliver successful outcomes to users and include the subsequent subcategories: product ownership, product + service add-on hardware, product + service add-on software, product availability, Product-as-a-Service (often shortened as PaaS), and product subscription.
- The data-driven business model is associated with the strategic dimension of customer intimacy. Furthermore, data-driven organizations adopt a Customer-to-Business (C2B) sales model, as businesses profit from the willingness of consumers to contribute data for the purpose of developing innovative digital-first products, thus fulfilling consumer needs. Data-driven business models originate from the digital operating model and comprise the following: indirect and direct data monetization, both subdivided into further categories.
- The platform business model is associated with the strategic dimension of ecosystem leadership. Platform-based businesses adopt a Customer-to-Customer (C2C) sales model, facilitating transactions between users, therefore allowing producers and consumers to locate each other, ultimately enabling organizations to maximize resource orchestration for the purpose of creating sustainable competitive advantage. Platforms presuppose interaction between participants through a third-party, mostly social media sites as well as e-commerce websites. Platform business models aim at generating value-creating interactions between external producers and consumers and comprise the following: data confederation, infrastructure community, social network, and marketplace [4].

Figure 3.2 graphically depicts the concepts briefly illustrated above.

Digital business models are characterized by digital business innovation and might be defined as models leveraging digital technologies in order to improve various aspects of organizations. Technological innovations enable businesses to enhance the value proposition as well as to deliver customized and meaningful experiences to consumers. Digitization entails the standardization of business processes. Furthermore, digital transformation across the manufacturing sector allows to achieve data-driven operational excellence, thus significantly lowering costs and

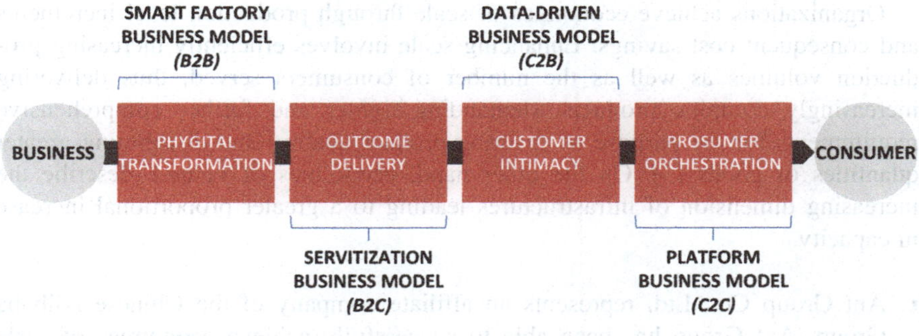

Fig. 3.2 Sales strategies and their associated business models

minimizing errors. Nonetheless, Jeanne Ross, organizational theorist and principal research scientist, argues that a clear distinction exists between digitized and digital companies. Digitization implies that organizations undergo a transformation aimed at enhancing traditional products as well as customer service. The principal benefits of digitization include the following: efficiency, operational excellence, and predictability. On the other hand, digital companies entail a rapid business innovation, thus contributing to delivering an innovative customer-centric value proposition. Successful organizations in the digital economy are going to be digital (providing customer value) as well as digitized (achieving scale and efficiencies). Becoming digital requires leaders to articulate a visionary digital value proposition in order to reassess the modalities through which digital technologies and information can enhance existing organizational processes, assets, and capabilities for the purpose of generating customer value. In addition, the digital transformation requires to reconsider the customer value proposition, thus prompting organizations to build a unique value proposition. Digital organizations invest in innovation for the purpose of designing and developing augmented products, providing services, and enhancing customer engagement. [5] Nonetheless, organizations must embrace digital transformation in order to meet the expectations of consumers, requiring digital-first goods and services capable of fulfilling customer demand. Companies must articulate a digital and unique value proposition, considering the role of recent technological advancements in delivering value to consumers. The widespread adoption of disruptive technologies within the organizational context enables companies to enhance their core businesses, mainly by increasing production efficiency through the introduction of digitized procedures. Nevertheless, the current industrial revolution is progressing from digitization to datafication, involving the generation and the collection of data, as well as the extraction of valuable knowledge.

Building efficient digital operating models requires the exploitation of economies of scale and volume, as well as economies of scope and complementarities and learning economies.

Organizations achieve economies of scale through production level increments and consequent cost savings. Enhancing scale involves efficiently increasing production volumes as well as the number of consumers served, thus delivering increasingly complex products (demanding tighter and further comprehensive requirements for the purpose of ensuring product quality) and distributing greater quantities of goods [3]. On the other hand, economies of volume describe the increasing dimension of infrastructures leading to a greater proportional increase in capacity.

- Ant Group Co., Ltd. represents an affiliate company of the Chinese Alibaba Group. Ant Group has been able to successfully achieve economies of scale and volume within the financial services sector. The corporation has managed to digitalize and automate processes as well as to implement artificial intelligence (AI) and machine-learning technologies. Human interactions have been limited and substituted by algorithms and further technological advancements. [3]

Economies of scope and complementarities occur during the simultaneous production of a wider variety of goods and services, thus resulting in lower marginal costs and increasing efficiencies. Economies of scope and complementarities represent cost advantages and figure internally within a particular organization or amongst several enterprises [3].

- The Eastman Kodak Company is an American technology company providing industry-leading hardware, software, consumables, and services for commercial print, packaging, and publishing. Following the shift from analog to digital photography, the company was not killed by the emergence of digital photography start-ups, instead it suffered financial losses due to the introduction of smartphones and social media platforms. Digital companies focus on connecting users as well as capturing and examining information flowing through their networks. Digital firms create as well as capture value in a different manner, therefore relying on a distinct operating model as opposed to Kodak's in order to deliver value to consumers [3]

The learning function of an operating model is considered to be essential to driving continuous improvement, enhancing operating performance over time, and developing innovative digital goods as well as services [3]. Leading companies are currently pursuing learning economies.

- Boston Dynamics is an American engineering and robotics design company founded as a spin-off from the Massachusetts Institute of Technology. Boston Dynamics is highly recognized for the development of an array of dynamic and highly mobile robots, including Spot. Spot represents an agile mobile robot, customized for a wide variety of applications, ranging from industrial inspection to entertainment. The quadruped robot serves to inspect dangerous, inaccessible, and remote environments, automate data collection on corporate websites and carry payloads on unstructured terrain. Autonomous and collaborative robots represent current innovative technological advancements. Collaborative robots, also referred to as cobots, constantly learn from human demonstrations, therefore fostering continuous improvement and performance.

Recently, the focus on learning and innovation has increased in order to deal with potential threats and capitalize on business growth opportunities. Digitalization and the consequent introduction of digital operating models has allowed emerging businesses to compete against incumbent (non-digitized or digitalized) organizations, therefore renovating established traditional sectors, thus pursuing economies of scale and volume, scope and complementarities, and learning.

3.1 Digital Operating Model

Operating models refer to the abstract and visual representation of the modalities through which organizations structure their core processes and deliver value to their customers. Operating models are not reserved for large corporations, regardless of size, all companies must articulate and implement an operating model in order to ensure strategy realization, business process optimization, as well as organizational alignment. Operating models range from simple to complex, depending on both organizational maturity and aspirations.

In a traditional business, size represents a doubtful advantage. As companies grow, businesses can deliver additional value at an economical price; however, the advantages of scale tend to be limited by the organizational operating model, thus encompassing assets and processes utilized for the purpose of delivering value to its customers. Organizations of significant sizes experience multiple issues due to their increasingly complex operating models. Ultimately, complexity becomes the downfall of traditional organizations, thus increasing operational costs as well as decreasing service levels.

Nonetheless, digital operating models embrace the advantages of digital scale, scope, and learning, in order to transform industries and reshape the nature of competitive advantage [3]. The emergence of digital operating models is transforming competition. Furthermore, digital operating models scale easily and continue to improve, despite the size and complexity of their operation [3]. Digital operating models are considered to be innovative as they tend to avoid direct human intervention during the good- or service-delivery process, therefore widely impacting the essence of an organization's operating model. 'The goal of a digital operating model is to deliver value at scale, to achieve sufficient scope and to respond to changes, engaging in sufficient learning' [3]. Constraints of any kind can be reduced as operational complexities are solved through software and analytics or outsourced to external partners within the network [3] (Fig. 3.3).

The exploitation of economies of scale is connected to consumer engagement as well as to the organizational monetization strategy, as 'the marginal cost of serving an additional user on multiple digital networks is zero (apart from the small

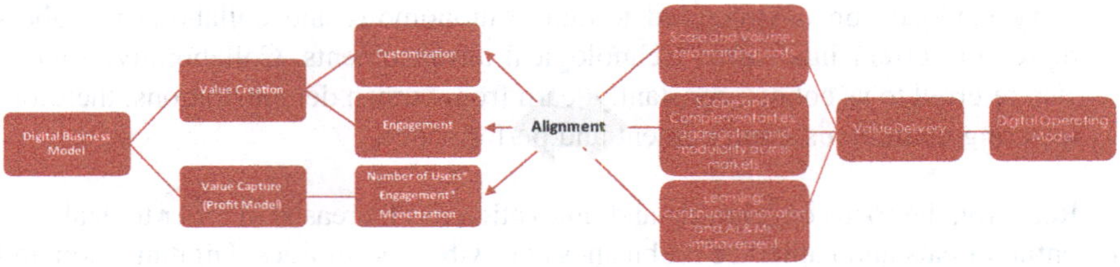

Fig. 3.3 Alignment between the digital business model and the digital operating model

incremental cost of computing capacity, easily available from cloud service providers)' [3]. Furthermore, monetization refers to the act of generating money from a particular product or service. In particular, data monetization is defined as the process of utilizing data in order to increase revenue.

Digital technologies are intrinsically modular and can easily enable business connections. Data become an organization's strategic asset, generating value through their cross-functional use within the enterprise. Sharing data externally contributes to generating increasing returns, eventually incrementing the competitiveness of the organization itself, as well as strengthening a firm's scope. Furthermore, economies of scope enable organizations to distribute a greater range of goods and respond to consumer behaviour shifts. Organizations achieve economic efficiencies through regular customer interactions. Customer engagement describes interactions between external consumers and organizations through multiple online as well as offline channels.

Subsequently, consumer data are gathered and elaborated, eventually leading to the production of customized goods or services. Digital business models can support enterprises in driving better engagement and interaction throughout the customer life cycle. In addition, digital operating models serve to support organizations in renewing the customer experience during the digital era. Therefore, adopting a digital-first mind-set enables organizations to redefine their relationship with customers.

Digitizing the operating model can also enable much faster learning and innovation. Continuous product–user interactions allow producers to collect useful data concerning consumer preferences and to design and develop customized products or services satisfying particular user needs. Once data are converted into information and knowledge, the value of learning increases alongside the organization's ability to comprehend its ecosystem and predict actions, thus potentially generating competitive advantages.

As organizations strive to deliver value and optimize scale, scope, and learning, operating models should match the direction set by business models. Over the years, scholars in operations strategy have argued that the actual performance of a business is optimized by the alignment between strategy and operations, namely between business model and operating model [3]. The digital operating model comprises multiple aspects as illustrated in the subsequent paragraphs.

3.1.1 Digital Operating Model: Product Dematerialization

The digital operating model is based on software development. Data analysts contribute to realizing software development, which utilizes algorithms in order to generate additional advanced algorithms, included within the digital infrastructure. Subsequently, data analysts are required to extract information from algorithms. Ultimately, cyber products are realized.

The term 'product dematerialization' describes the essence of the current digital trend as well as the principal feature characterizing cyber products. Cyber products represent immaterial goods available to users. Digital economies arise as a consequence of the production of cyber goods or services.

Figure 3.4 illustrates the digital operating model, with reference to product dematerialization, within the business model canvas for the purpose of highlighting its effects on the various building blocks.

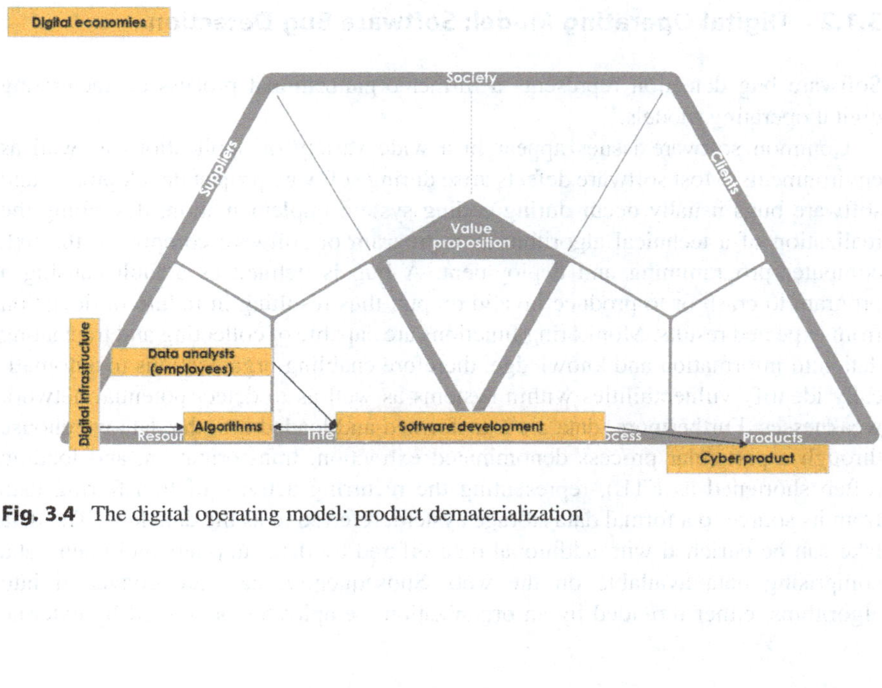

Fig. 3.4 The digital operating model: product dematerialization

3.1.1 Digital Operating Model: Product Dematerialization Google LLC

G Suite represents an example of cyber product created through the process formerly described. Google Workspace (formerly referred to as G Suite) features a collection of cloud computing, productivity and collaboration tools, software, as well as products developed and marketed by Google. Most of its services are individually available at no cost to consumers, who are guaranteed access to all services through their free Google (Gmail) accounts.

3.1.2 Digital Operating Model: Software Bug Detection

Software bug detection represents a further organizational process characterizing digital operating models.

Common software issues appear in a wide variety of applications as well as environments. Most software defects arise during software project development, and software bugs usually occur during coding system implementation, describing the realization of a technical algorithm as a program or software component through computer programming and deployment. A bug is defined as a fault causing a program to crash or to produce invalid output, thus resulting in failure or deviation from expected results. Monitoring functions are capable of collecting and translating data into information and knowledge, therefore enabling organizations to automatically identify vulnerabilities within systems as well as to detect potential network weaknesses. Furthermore, data are transformed and loaded into the data warehouse through a particular process denominated extraction, transformation, and loading (often shortened as ETL), representing the recurring activity of transferring data from its source to a formal data storage system, referred to as the data lake. The data lake can be enriched with additional data offered by data suppliers and open data, comprising data available on the web. Subsequently, data are converted into algorithms, either provided by an organization's employees or offered by external

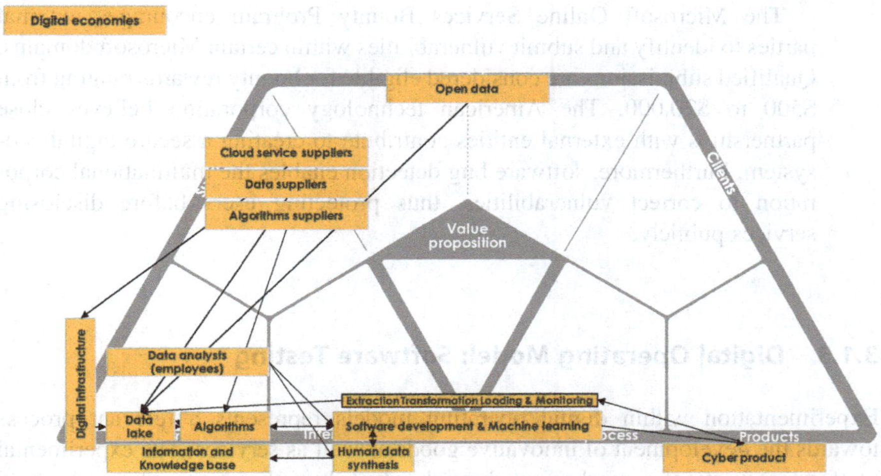

Fig. 3.5 The digital operating model: software bug detection

providers, including algorithm suppliers. Algorithms can be included within multiple software, whilst machine-learning techniques serve to discover patterns in data, which are utilized to formulate predictions regarding further data points. Automated software enables organizations to effectively discover faults within the software development process, affecting program functionality and causing incorrect results and to run critical systems and operations, while modernizing information technology and optimizing data architecture.

Figure 3.5 illustrates the process of resolving bugs within software as well as computer programs, within the business model canvas for the purpose of highlighting its effects on the various building blocks.

3.1.2 Digital Operating Model: Software Bug Detection Microsoft Corporation

The Microsoft Online Services Bounty Program encourages external parties to identify and submit vulnerabilities within certain Microsoft domains. Qualified submissions are considered eligible for bounty rewards ranging from $500 to $20,000. The American technology corporation believes close partnerships with external entities contribute to creating a secure digital ecosystem. Furthermore, software bug detection enables the multinational corporation to correct vulnerabilities, thus protecting users before disclosing services publicly.

3.1.3 Digital Operating Model: Software Testing

Experimentation within digital operating models represents a relevant process towards the development of innovative goods as well as services. The experimental platform consists of a mechanism through which hypotheses regarding advanced prediction and decision algorithms are tested in order to ensure the changes suggested are generating the intended causal effect. An innovative experimental platform is going to provide the comprehensive set of technologies, tools, as well as methods required to perform experimentation at scale.

In order to utilize an experimental platform, potential significant changes to the business must first be formalized as a hypothesis. Each hypothesis is going to be tested as a randomized control trial (often shortened as A/B test) consisting of a random sample of users, who are going to be exposed to the change (referred to as treatment) and a second random sample of users, who are going to experience business as usual (the control). The outcomes are later compared, and in case the difference between them is statistically significant, the treatment is going to actually affect the outcome, instead of simply being spuriously correlated. The aforementioned approach is represented within the illustration below (Fig 3.6) and aims at ensuring any prediction being generated by algorithms actually produces a causal effect on the outcome [3]. Further test procedures include fail-fast testing, encouraging software developers to fail fast and early, forcing software to fail immediately in order not to postpone failures. Fail-fast enables to stabilize software instantly and to significantly reduce the cost of faults as well as of software defects. Furthermore, the aforementioned test procedure enables software bugs to be detected earlier and fixed within a short period of time. Additional techniques worthy of mention include the 'release now, patch later' approach, typically associated with the video game industry. Patch management refers to the process of distributing and applying updates to software. Patches are often necessary to correct errors, also referred to as bugs or vulnerabilities, within software. Common areas necessitating patches include operating systems, applications, as well as embedded systems. Once vulnerabilities have been identified, following the release of a particular software component, a patch can be utilized in order to fix the faults detected, thus ensuring assets are not susceptible to exploitation. The digital economy is characterized by quick experimentation. Businesses are embracing digital experimentation for the

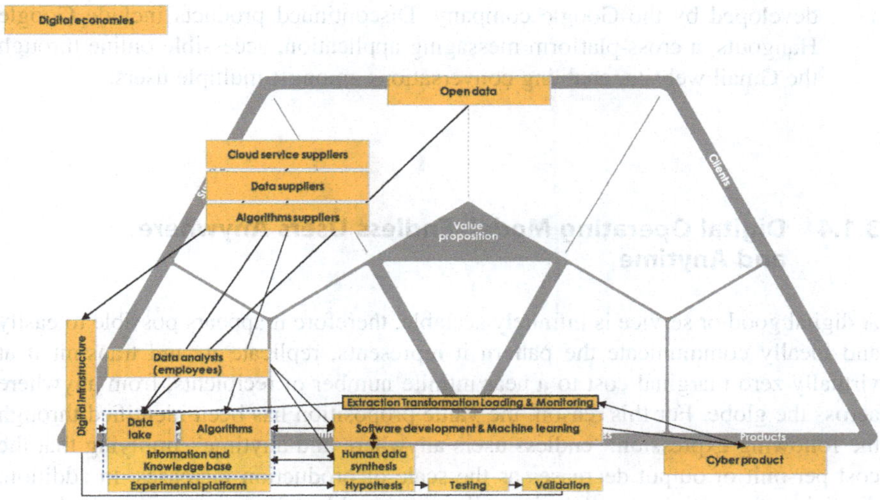

Fig. 3.6 The digital operating model: software testing

purpose of accurately targeting marketing campaigns and of testing applications in real-time and at large scale. Artificial intelligence (AI) offers organizations the principal tools and techniques in order to accelerate the experimental practice.

Figure 3.6 illustrates software testing within the business model canvas for the purpose of highlighting its effects on the various building blocks.

► 3.1.3 Digital Operating Model: Software Testing Google LLC

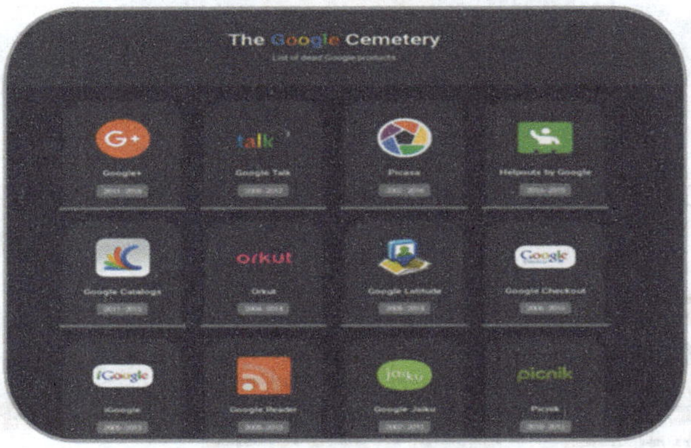

Google LLC experiments on multiple projects, nonetheless some fail. Google Cemetery constitutes an Internet website and a digital graveyard comprising a list of discontinued applications originally designed and developed by the Google company. Discontinued products include Google Hangouts, a cross-platform messaging application, accessible online through the Gmail website, enabling conversations amongst multiple users.

3.1.4 Digital Operating Model: Endless Users Anywhere and Anytime

A digital good or service is infinitely scalable, therefore it appears possible to easily and ideally communicate the pattern it represents, replicate it, and transmit it at virtually zero marginal cost to a near infinite number of recipients, from anywhere across the globe. For this reason, the value proposition has been identified through the following expression: 'endless users anywhere and anytime', implying that the cost per unit of output decreases as the scale of production increases. In addition, digitizing the activity renders it easily connectable, also at zero marginal cost, dramatically increasing its scope [3].

Figure 3.7 contributes to introducing the value proposition characterizing innovative digital operating models.

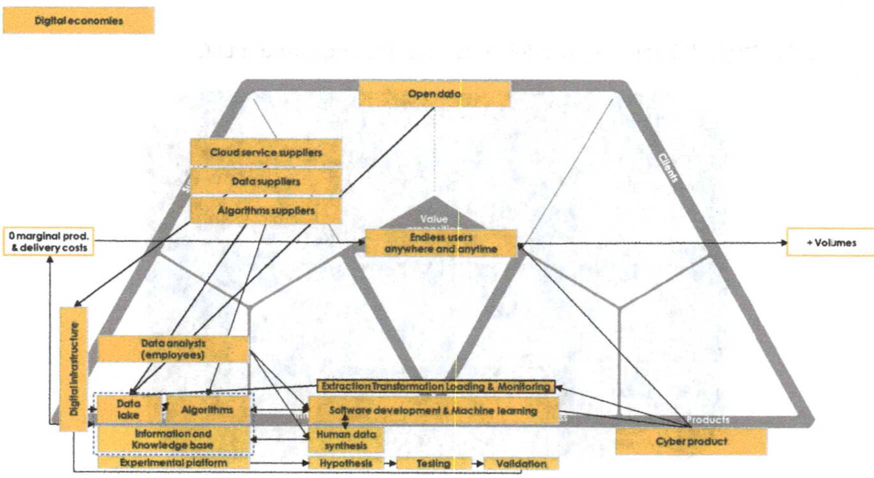

Fig. 3.7 The digital operating model: endless users anywhere and anytime

3.1.4 Digital Operating Model: Endless Users Anywhere and Anytime Google LLC

The Google application consists of a free online map, thus constituting an accurate example of cyber product and representing the concept of product dematerialization, as opposed to traditional paper road maps. Google Maps improves its functionality based on the indirect contribution of its users. In particular, data collection and analysis appear essential to delivering value to customers, facilitating outcomes, as the application incorporates data within the choice of the shortest routes, otherwise unavailable.

3.1.5 Digital Operating Model: Non-Traditional Revenue Model

Ultimately, companies reach customers, who purchase cyber products, thus producing transactional revenues. Purchasing a cyber good or service does not affect product ownership, as users acquire a software license in order to authorize one computer program to be utilized on multiple computers or by a wide number of users. Therefore, organizations retain ownership of the actual dematerialized good.

Figure 3.8 illustrates the sales process characterizing digital operating models.

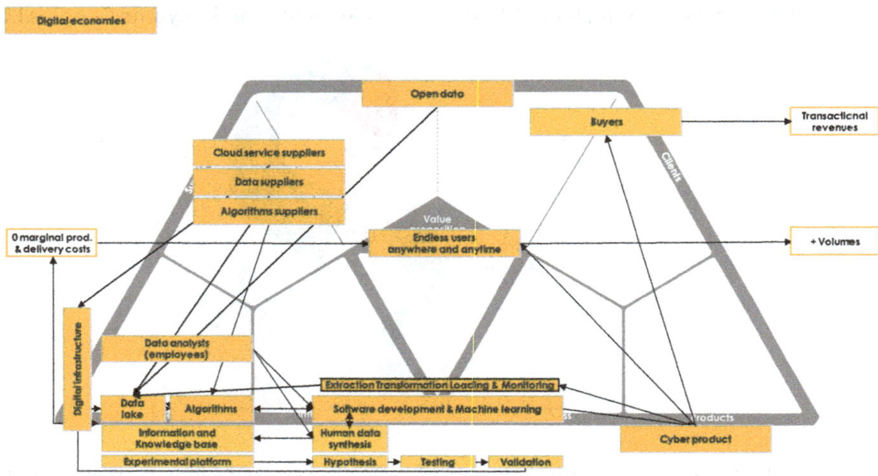

Fig. 3.8 The digital operating model: non-traditional revenue model

3.1.5 Digital Operating Model: Non-Traditional Revenue Model Microsoft Corporation

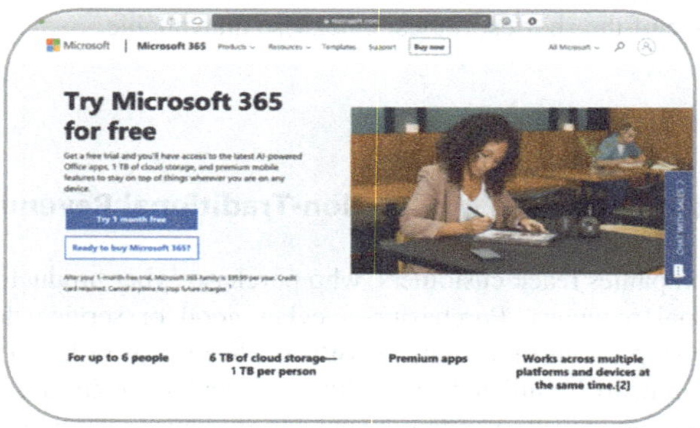

Microsoft sells a static digital service to a potentially infinite customer base. Microsoft 365 is sold as a one-time purchase, namely users pay a single, up-front cost to obtain Office applications for one computer. Microsoft offers a non-exclusive license, authorizing businesses as well as individuals to utilize the product or service in return for a fee. Thus, users agree to the terms requested by Microsoft and purchase a license to utilize the software.

3.2 The Introduction of Artificial Intelligence within Digital-First Products

Amongst the major disruptive technologies driving Industry 4.0, artificial intelligence (AI) is worth mentioning, particularly with respect to its implications on the digital operating model and on the further business models which have already been identified and briefly discussed at the beginning of this chapter. The specific term 'artificial intelligence' was coined by Professor John McCarthy during the summer of 1956, when he held the first academic conference on the subject at Dartmouth College. Following his view, machines would have to be initially equipped with rules of logical reasoning and real-world conditions and trained to follow and react to those same instructions [6]. Nonetheless, this approach resulted time-consuming for programmers, and its effectiveness relied thoroughly on the clarity of rules and definitions. Modern artificial intelligence (AI) has deviated from the aforementioned approach by adopting the notion of machine learning. In particular, artificial intelligence (AI) consists of machines exhibiting intelligence, as characterized by sufficient computational resources and algorithms defining a learning process, for the purpose of acquiring knowledge as well as of enhancing performance [7].

Artificial intelligence (AI) is increasingly gaining prominence in multiple industries across the world. Artificial intelligence (AI) is transforming companies, innovating the modalities through which businesses operate and navigate the global marketplace. Recently, artificial intelligence (AI) has contributed to shaping modern society, primarily augmenting human capabilities. Businesses driven by artificial intelligence (AI) deliver value to customers through the innovative utilization of software instructions and algorithms, thus enabling an organization to shape, deliver, and drive value effectively. In accordance with the literature, artificial intelligence (AI) represents the advanced operational foundation of business, namely the heart of an organization's operating model, defining the system through which companies structure their core processes and manage the execution of tasks [3]. Furthermore, artificial intelligence (AI) (including its deeper concepts) intends to revolutionize the current organizational structure, as presently, traditional businesses have considered evolving their underlying business models, thus proactively adapting to the frequently shifting technology landscape. Organizations seem to favour the transition towards a significantly automated and sustainable framework, thereby successfully embracing change in order to drive innovation.

In addition, companies benefit from the numerous advantages deriving from the implementation of AI-driven processes as they appear to be highly scalable compared to traditional processes, enabling greater scope by connecting with multiple digital businesses, therefore contributing to the creation of opportunities for learning and improvement [3]. The Fourth Industrial Revolution is offering companies the chance to transform their manufacturing operations and to manage relationships within the marketplace differently than in the past [8]. Technology represents assuredly a valuable ally in building powerful connections with business partners and customers for the purpose of recognizing and grasping profitable business opportunities as well as of driving revenue growth. Nowadays, a growing number

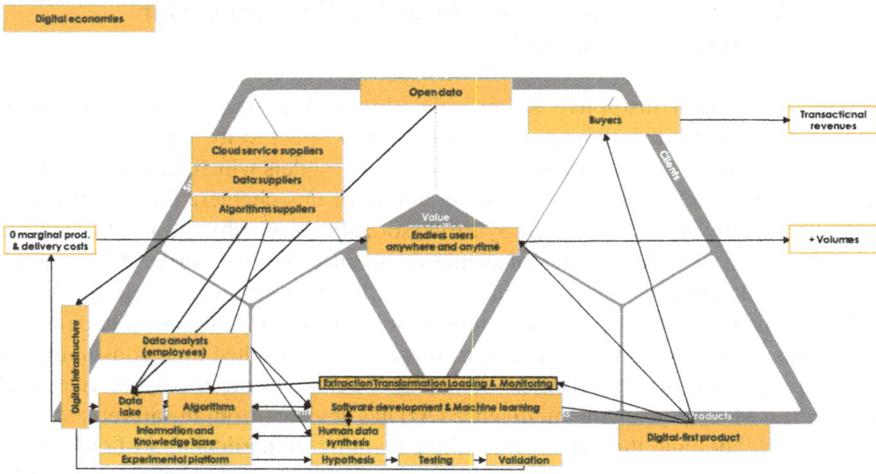

Fig. 3.9 The digital-first product

of organizations are involved in the technology business, considering the relevance of technology in day-to-day operations. In particular, digital technology is key to future business success as technologies are based on the collection and utilization of information in the form of data.

Currently, data represent the primary asset available to companies; specifically, data are considered a strategic asset to organizations, as all kinds of analysis start from data. Furthermore, data are central in guiding business leaders to take informed decisions, although attention must be paid to the enormous outflow of data as well as of information, which typically confuses leaders in identifying the appropriate information, suitable for the development of innovative strategies.

Digital technology refers to a process consisting of four principal activities: data collection, data transmission, data elaboration, and 'data learning', defining a learning function through which digital technology is expected to act based on temporarily evolving criteria depending on feedback. A software, namely an algorithm characterized by a clearly defined sequence of operations producing an output, is employed in performing the aforementioned practical steps [8].

Certainly, technology is applied widely in multiple industries as organizations are becoming accustomed to automating business processes. Nonetheless, technology must not be incorporated exclusively into a firm's daily operations, instead it should be extended to its products and services as well.

Figure 3.9 introduces the concept of digital-first product within the business model canvas.

A digital-first product consists of a good whose functions are controlled by software. Nevertheless, software must not necessarily be included physically within the particular product or service, it appears sufficient to equip the determined good with surface sensors enabling direct connection to software and communication with additional surrounding items. Augmented products are characterized by the effective integration of smart sensors as well as actuators, transmitting enormous quantities of data. As a result of the purchase and the consequent utilization of the digital good or service, consumers provide companies with a deeper understanding of user needs, preferences, as well as desires.

Computers, applications, and smartphones are a few examples of digital-first products. Nonetheless, there are further consumer goods whose functions are traditionally governed by analogue mechanisms and eventually assigned to software. Automobiles and household appliances include examples of traditional products converted into digital-first goods. Essentially, technology can be integrated within any product or service, irrespective of its nature. Food and medicines could be controlled by software, monitoring their functions and comprehending their utilization [8]. Although most companies already incorporate software within products, they do so in order to enhance the functions associated with the specific good or service. In this regard, organizations do not revolutionize their underlying business models [8]. Properly incorporating software within products implies to reconsider product innovation and to convert the product into a useful instrument, allowing the completion of the following actions characterizing digital technology: collecting, transmitting, and processing data, as well as executing actions as a result of data comprehension. For this reason, companies must focus on generating augmented products, namely goods equipped with artificial intelligence (AI) optimizing their functionalities and connecting with users, thus generating value. During usage, products generate on-going benefits for both consumers and producers, as goods may transmit valuable data to producers [8].

The digital revolution requires to position products at the centre of the relationship between enterprises and customers. Digital-first products become the primary protagonists, as augmented goods contribute to generating innovative business models as well as to building advanced value ecosystems, enabling constant communication with users, for the purpose of discovering relevant and previously unidentified benefits. Continuous product–user interactions give rise to an entirely innovative approach, defined as the customer-to-business (C2B) perspective and involving a proper shift, significantly impacting on both business models and operational processes [8].

Interestingly, customers are not intrigued by technology per se, instead consumers value its potential, namely the advantages offered by technological advancements, especially in terms of searching for a solution to a problem. In this respect, Parrot's initiative is worthy of mention. Parrot SA is a wireless products manufacturer company headquartered in Paris, France, and characterized by a visionary approach to the user experience design. The organization's smart product portfolio includes a connected pot comprising a smart watering system and four sensors, continuously monitoring the plant's conditions: light sensors, in order to evaluate the light exposure throughout the day; temperature sensors, so as to assess the environmental conditions; moisture sensors, to determine the necessity of watering the plant and chemicals sensors, for the purpose of estimating the effects of fertilizers on plant growth. Data collected through the smart pot is directly transmitted to the corresponding application (previously downloaded to the user's smartphone), which includes a comprehensive database containing information on the growing conditions of a multitude of plants. Subsequently, users are requested to insert the required information within the app and simply follow the instructions provided by the relative application. Therefore, the plant caretaker is offered the opportunity to acquire the necessary instant skills in order to garden successfully, while being completely supported by technology throughout the entire process. In this particular situation, digital technology fosters learning, providing users with the necessary competencies in order to perform the appropriate tasks. The smart pot constitutes an example of reverse engineering, a concept describing the requirement of consumers to defer learning-related efforts to technology in order to fully benefit from the acquisition of a new skill [8]. The previous considerations appear significant as they determine the main factors driving consumers to purchase a good or service in the near future; said otherwise, being able to identify this kind of information allows organizations to better understand the drivers which are going to profoundly influence customer purchase choices as well as decisions.

Nonetheless, the development of digital-first goods or services requires organizations to investigate the effects of technologies applied to products on multiple levels; these include the following: personal, social, business, and planet.

- The personal level focuses on the benefits brought by technologies to individuals' lives.
- The social level explores the impacts of technological innovation on society and on social relationships.
- The business level considers the consequences of technological change within the industrial sphere.
- The planet level examines the environmental implications of technologies on the planet, including its resources [8]

Traditionally, products are designed to adequately perform a certain function. Once the product is completed, it is distributed to the customer for the purpose of utilizing the determined good within a particular context. Nonetheless, following this perspective, companies tend to release products which are considered to be obsolete, although demonstrating their promising performance, as they lack specific technologies enabling the evolution of the user experience in real-time, through data [8].

Digital-first products are engineered in order to adapt to consumer habits and to accurately anticipate user needs. Therefore, organizations must provide digital goods as well as services carefully designed according to two key concepts: connection and progress. The former refers to establishing a proper connection with users, collecting data during product utilization, for the purpose of elevating innovation and enhancing the consumer experience, as well as creating a sustainable customer relationship. The latter consists of introducing users to ideas and services they were previously unaware of, as customers of the future are willing to interact with companies satisfying a need which cannot be filled anywhere else. Connection and progress are necessarily intertwined and are crucial to the development of innovative goods and services, ultimately, guaranteeing enterprises the possibility to capitalize on emerging technologies as well as on business growth opportunities [9].

The digital era is characterized by two major aspects: data utilization and customer centricity. Data are a necessary and central element in the development of goods and services. Data feed software, responsible for the correct functioning of the digital product. Nonetheless, data can be thought of as fluctuating knowledge, which the digital-first product captures and elaborates, allowing a digital good to evolve through the evolution of the software controlling it [8]. Subsequently, customer-related data are gathered and meticulously analysed with the intention to considerably increase product benefits. In this sense, data serve as a valuable instrument providing an understanding of customer requirements as well as preferences, in order to properly satisfy consumer needs.

Digital-first products exhibit eight distinct characteristics.

1. First, digital goods comprise the ability to predict the future and to proactively react to complex problems or to optimize operating systems. Furthermore, augmented goods can be designed in order to anticipate product usage: their predictive ability is based on the capability of processing information for the purpose of building scenarios, assigning a probability of occurrence to each outcome.
   Most companies have already enabled predictive maintenance, a proactive strategy utilizing condition monitoring tools as well as data analytics to detect deterioration signs and component failure in order to conveniently schedule corrective maintenance.
2. Second, digital-first products are considered to be prescriptive as digital goods as well as services convert information, generally obtained during product utilization, into advice, suggestions, as well as instructions for the purpose of offering users personalized recommendations, thus enhancing the consumer experience.
   Over the years, the global beauty industry has experienced disruptive innovations, releasing smart skincare solutions. L'Oréal recently launched an artificial intelligence-powered digital skin diagnostic tool, utilizing selfies in order to provide accurate and personalized treatments, as well as skincare recommendations, therefore delivering an authentic experience to consumers. L'Oréal Paris Skin Genius evaluates five core skin attributes, ultimately suggesting a tailored routine of skincare products better suited to age-related changes of skin, thus greatly improving the ability to battle the effects of aging on appearance.
3. Personalization of digital-first products implies the collection of data for the purpose of connecting user preferences and behaviours to product development. Equinox Group, an American luxury fitness chain, has created its own mobile application providing an entirely digital training experience. The Equinox application is designed to support members in structuring proper custom workout routines, depending on users' precise fitness and health goals, physiques, as well as their current state of mind.

4. In addition, digital-first products differentiate themselves from traditional products, as digital goods are upgradable. Software controlling digital products are continuously upgraded to enhance and increase product performance.
   The automotive industry is undergoing a profound transformation driven by digital technologies and business models disruptions. Increasing customer demand for highly connected autonomous and electric vehicles has led car manufacturers around the world to reimagine the future of transportation. Tesla, Inc. is an American electric vehicle and clean energy company headquartered in California, producing vehicles by developing software on unique hardware. Tesla is an automobile controlled by a constantly evolving software. Periodically, the parent company releases over-the-air software updates adding new features as well as enhancing existing functionalities via Internet connection. All Tesla vehicles are designed in order to improve over time through regular software updates.
5. Furthermore, digital-first products interact with further systems and devices, structuring high-level network protocols connecting numerous systems, based on distinct platforms. Digital goods are designed to become platforms, within which interfaces are enabled in order to allow third parties to exploit resources and cooperate towards the creation of emerging business models and the generation of sustainable value. The potential behind data sharing has contributed to the rise of a specific phenomenon referred to as API economy. APIs, namely application programmer interfaces, constitute software components created for the purpose of enabling communication amongst platforms, applications, as well as digital systems, therefore properly exchanging information and data, with the ultimate goal being the conversion of digital goods into third-party revenue sources.
   Amongst the major leading players in the ridesharing market, Uber represents the most popular ridesharing service currently operating. Uber holds an astonishing amount of anonymized and aggregated data about its users. In particular, based on information collected over a vast period of time, Uber has recently developed a tool, sharing the dynamic insights about traffic and mobility in those cities within which it operates. Uber Movement displays average travel times around and between several areas of available cities. The current initiative intends to support urban planners in taking informed decisions regarding cities and existing infrastructures and investing in future solutions in order to ensure safe as well as efficient mobility. Furthermore, the American technology enterprise shares considerable amounts of data with a wide variety of organizations. Interestingly, data gathered revealed that Uber riders typically spend around 2 billion dollars a year in their local communities, often, shortly after stepping out of an Uber. In this regard, Uber has partnered with Visa in order to support local businesses, introducing the 'Local Offers' Programme, through which merchants can encourage members to shop and dine in their stores and restaurants in exchange for Uber points, driving increased customer acquisition and engagement, as well as extending the rider experience beyond the car and into local communities. Uber Local Offers is powered by the Visa Commerce Network, whose platform

connects transactions between two merchants; whilst cardholders are seamlessly rewarded for their daily purchases, local businesses benefit from customer retention.

6. Designing a digital-first experience requires creating a hassle-free customer journey, guaranteeing constant user–product or service interactions.
   Recently, online shopping has become increasingly popular for a wide variety of reasons, as opposed to in-store shopping. Despite the coronavirus pandemic, fashion retailers have soared, reporting significant growth in sales as well as in profits. Amazon has launched Prime Wardrobe, an exclusive programme offering customers the possibility to try on the preferred clothing items before completing their purchase. Members have one week to try on the items at home and are going to be charged for the items they decide to keep. Initially, the service featured the introduction of an Alexa-enabled smart speaker for fashion advice: the Echo Look camera, a virtual fashion assistant evaluating current clothing trends and providing fashionable recommendations through machine-learning algorithms and human fashion experts. Originally, users would take a picture of their outfits, and the application would suggest style advice; nevertheless, the company has recently decided to discontinue its Alexa device.
7. A digital-first product favours disintermediation, creating a direct relationship between the counterparts involved in the process.
   Removing intermediaries notably improves the e-commerce supply chain. In particular, Amazon has launched 'Part Finder', a feature embedded within the Amazon application and built utilizing the technology acquired from Partpic, representing an American visual recognition technology start-up. The present advanced solution consists of a product-finding function through which consumers can research products without entering text. The phone camera identifies the object in question, so the application can scan it, match it, and direct customers to corresponding items from Amazon's digital product catalogue.
8. Finally, digital-first products are endowed with artificial intelligence (AI), enabling innovative goods to learn, adjust, and continually evolve. Furthermore, digital devices learn from their surroundings, as their development is accelerated through communication amongst each other. The ability to transfer learning is definitely powerful and impressive. Transfer learning has been identified as a deep learning neural network technique, transferring knowledge from a pre-trained model used to solve a certain task and applying it to a similar problem within the same domain [10]. In addition, connected learning stores knowledge within a machine-learning model, thereby drastically reducing costs in order to build training models, reaching economies of scale through artificial intelligence (AI).
   Technology is certainly moulding the toy industry, as manufacturing units can now deploy machine-learning techniques as well as artificial intelligence (AI) technologies, thus renovating the domestic play industry. In modern toys, it is possible to notice a considerable level of programmed intelligence, which can be utilized to control their actions or prompt them to respond to input in order to

provide real-time feedback as well as interaction. A prime example in this particular field is represented by Cozmo, an endearing and sophisticated palm-sized robot, powered by artificial intelligence (AI), whose abilities and potential develop constantly through frequent application updates in order to provide innovative methods of play. Cozmo represents an educational toy, teaching children the basics of coding; also, its advanced facial recognition software allows it to recognize faces as well as to pronounce names [8].

Digital products enable innovative value propositions as well as advanced business models. The diffused implementation of sensors and actuators within digital products, enabling organizations to extract data and to gain valuable product-user insights, represents a phenomenon referred to as prediction economy. Prediction economy enables the creation of innovative business models as well as the introduction of subscriptions and pay-per-use pricing models. Products are going to be conceived in 2D and 3D simulations, generated by advanced manufacturing tools, through processes powered by digital printing as well as laser cutting, in order to achieve seamless manufacturing processes from design to product at an unprecedented speed. Furthermore, leading organizations understand that customer retention and loyalty are achieved by delivering inspiring consumer experiences. The digital culture expects the industrial designer's role to morph considerably as the creative mind-set evolves the designer from operator to strategic thinker. As mentioned by Tim O'Reilly, industrial designers must connect with customers for the purpose of creating digital goods capable of satisfying consumer expectations, while continuously enhancing products through a process referred to as Learn-Build-Measure in order to deliver an outstanding user experience [11].

An additional emerging theme alongside artificial intelligence (AI) is certainly represented by the Internet of Things (IoT). The term refers to the application of the latest digital technologies to product development for the purpose of designing an artificial intelligence-enabled entirely autonomous object performing the following actions: sense, connect, and think [8].

- Sense. The ability of a digital device to sense its surroundings is determined by the object's capability to autonomously collect information through sensors incorporated within the augmented product. Cameras are amongst the most common and widely employed typologies of sensors presently available on the market.
- Connect. The capability of a digital item to connect to further devices refers to the product's ability to transmit data previously collected from the object itself. Whenever digital products are connected, the availability of data increases exponentially, as during product usage, objects interact with each other transferring valuable data regarding their utilization. The higher the number of augmented products connected, the greater the volume of data and the consequent possibility to understand the underlying system.
- Think. The third and last step of the process formerly described includes the digital product's ability to elaborate data, acquiring knowledge as well as

predictability. Processing algorithms influence the results obtained by the previously gathered and transmitted data.

Artificial intelligence (AI) represents a useful input in decision-making. Nevertheless, human judgement remains essential to navigating and untangling the decision-making process within organizations. Paul Michelman, editor-in-chief of the MIT Sloan Management Review, argues whether algorithms are going to affect organizational strategies. Although individuals choose to follow human judgement rather than relying on mathematical instructions delivering superior results, machine-learning techniques serve to categorize and organize data as well as to identify patterns within datasets for the purpose of enhancing and formulating predictions. Nonetheless, emotional and intellectual considerations allow to convert information into business advice [11].

3.3 Data-Driven Business Models

Data-driven business models mainly impact the following building blocks: resources, customers, suppliers, internal processes, external processes, and products. Datafication is characterized by the exploitation of enormous amounts of data available to companies, for purposes other than those for which they were generated. Therefore, data contribute to digitizing processes, products, and customer relationships in order to increase the volume of organizational data. Data also contribute to generating additional revenue streams through the sale or exchange of corporate data for the purpose of improving processes and products as well as the customer experience. In indirect and direct data monetization, data provide organizations with powerful insights about individual customers. For these reasons, data-driven business models are associated with the strategic dimension of customer intimacy [12–19]. Customer intimacy occurs once customer needs are fulfilled through the sale of personalized products that match the unique needs of a niche. It is based on a value proposition which considers the following critical success factors: product range and customization, complementary goods, services and experiences, as well as greater customer involvement and sharing [20, 21].

The data-driven business model recognizes that the information capacity of a single datum is superior to that which justified its generation. The aggregation of data and therefore of the information available can generate knowledge in areas that differ from its origin. The possibility for a company to adopt a data-driven business model depends on the volume of data available, but also on their speed, variety, value, and truthfulness, hence it appears important for organizations to invest in:

1. data generation, for the purpose of creating the largest amount of structured organizational data;
2. data collection, for the purpose of collecting data available on the inside and on the outside in an effective and efficient manner;

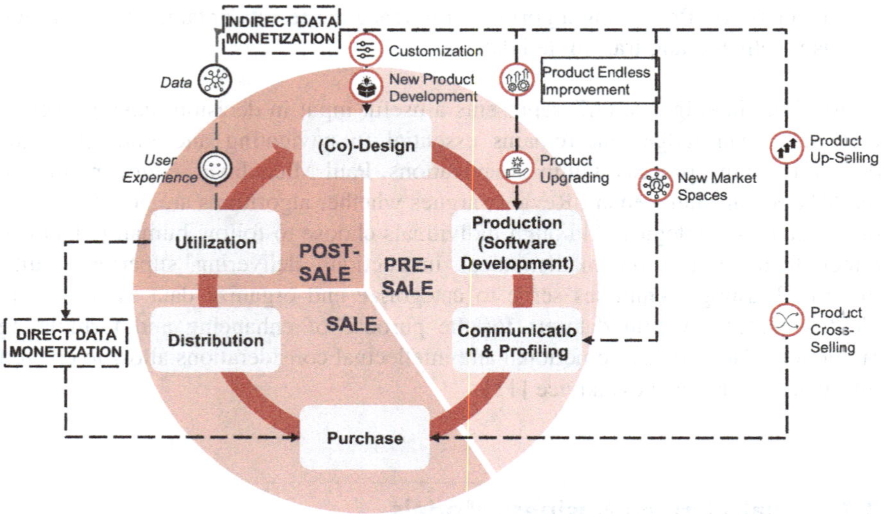

Fig. 3.10 The data-driven business model

3. data integration, for the purpose of correlating the data collected in order to contextualize data, and therefore transform data into information;
4. data visualization, for the purpose of translating information into a visual context; and
5. data analysis, for the purpose of processing the information selected in order to transform it into knowledge to improve the decision-making skills [22–26].

The data-driven business model represents an emerging digital business model, developed and evolved from the aforementioned digital operating model. Data-driven business models originate from the immense quantities of data arising as a consequence of the introduction, within organizations, of multiple technological innovations characterizing the Fourth Industrial Revolution. Considering the relevance of data in the digital era and the broad technological and economic shifts contributing to favouring the rise of software companies, the data-driven business model represents the underlying logic driving organizational strategy as well as the basis for all the further business model categories identified.

Internal and external data represent a valuable source of information for organizations. Data are continuously exploited for the purpose of improving production efficiency, anticipating maintenance issues, developing innovative digital-first products, or enhancing the performance of existing goods and services.

The data-driven business model can be schematically illustrated through the diagram pictured in Fig. 3.10.

Data utilization allows extraction of relevant information, gaining insights into consumer behaviour patterns in order to enhance the user experience. Data analysis favours product endless improvement as well as new product development.

Furthermore, evaluating user–product interactions enables organizations to enhance the most appreciated features of digital-first products and design updates for the purpose of introducing advanced versions of determined digital goods and services. Updating a product's software, introducing additional functionalities as well as features, depending on product usage data, contributes to enriching the consumer experience, therefore converting consumer insights into actionable results. The highest-performing and fastest-growing companies have adopted data monetization and have incorporated it within their organizational strategy. Furthermore, data are typically utilized to discover additional business opportunities and to reach multiple customers.

Embedded analytics seamlessly integrate analytic capabilities and content into digital goods or services, providing relevant advantages for producers as well as consumers. The customer journey comprises two critical stages: the moment of purchase and the post-sale process. Customer needs appear the motive prompting consumers to purchase a determined product or service. Once consumer needs have been identified, organizations must fulfil the requests, offering individuals digital goods addressing their requirements. The post-sale process is alternatively characterized by the acquisition and utilization of augmented products and services. Nonetheless, whether the buyer intends to repurchase the digital product strictly depends on a particular factor referred to as customer loyalty.

Furthermore, data-driven business models have contributed to reversing the sales process models dominating the past century. Modern-day sales processes have changed dramatically due to the current technological trends, namely connectivity, intelligence, as well as flexible automation, driving the Fourth Industrial Revolution. Business-to- Consumer (B2C) sales have been substituted by Consumer-to-Business (C2B) sales models. The concept refers to a particular economic relationship through which consumers contribute to generating value, whilst organizations consume the value created. Consumer-to-Business (C2B) practices are facilitated by the emergence of innovative technological advancements. The rise of the Consumer-to-Business scheme is partly due to the decreasing costs of technology as well as to the advent of big data analytics, enabling businesses to extract value from consumers and to gain a better understanding of them. The Consumer-to-Business perspective is represented in Fig. 3.11.

Product utilization allows collection of data as well as information which are transformed into knowledge through the process of data elaboration, enabling organizations to gain valuable consumer insights. Knowledge serves to develop the value proposition. Nevertheless, organizations must focus on their production processes either adjusting the product, upgrading the software with a latest version, or designing innovative goods or services, resorting to strategies such as up-selling in order to increase product value as well as consumer value, thus growing recurring revenue.

The data-driven business model is further classified into two distinct categories: indirect and direct data monetization, further subdivided into additional sub-categories, including product endless improvement, customization, product

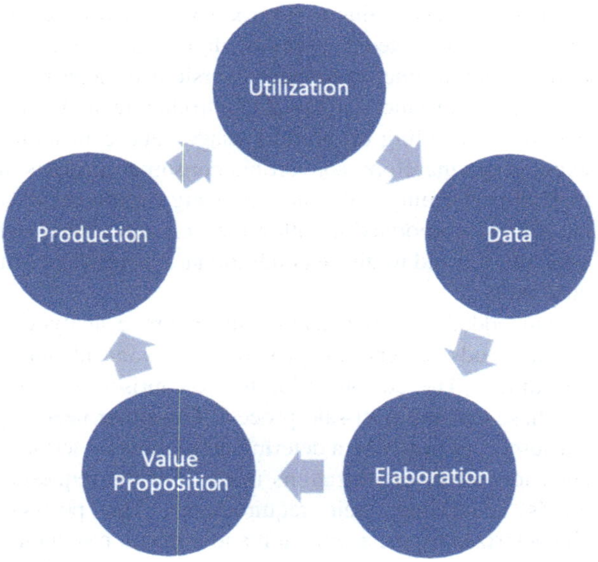

Fig. 3.11 The Consumer-to-Business (C2B) sales model

upgrading/up-selling, new product development/cross-selling, and new market space creation.

The following paragraphs delineate the implementation of the aforementioned monetization strategies.

3.3.1 Indirect Data Monetization: Product Endless Improvement

Indirect data monetization refers to the process of analysing and processing data available within the company for the purpose of extracting value. The aforementioned data enhancement process requires companies to integrate data from two distinct sources: products and users. User-generated data derive from online feedbacks, comments, and video or audio content. Alternatively, product-generated data are obtained from the features deriving from the product itself. Nonetheless, data can also be classified according to the individual producing them, including users, products, manufacturers, or service providers. In this case, we can distinguish among six distinct kinds of data: active data generated by users and deriving from voluntary inputs, passive data generated by users and deriving from involuntary inputs, data collected from products, data produced by manufacturers, and data generated by the third-party service provider (service updates). Data also originate from software such as ERP and CRM, from digital platforms or from intelligent devices equipped with IoT sensors. Integrating data requires building a trusted data marketplace where those rendering data available, through the process of direct data monetization, and those acquiring data, through the process of indirect data monetization, can build trust in the relationship. The secure circulation of data represents a fundamental pillar in order to build a data-driven economy. The data held and acquired will contribute to improving corporate skills and will be available to the customers in the form of an enhanced and personalized customer experience.

Internal or indirect methods include utilizing data to perform measurable business performance improvements and to inform decisions. Furthermore, organizations collect data from multiple strategic parties within the supply chain and store information within data warehouses for the following purposes:

- Conducting industry analysis, thus gathering and comparing quantitative as well as qualitative information in order to facilitate an organization's understanding of its competitive position relative to additional companies producing similar augmented products.
- Monitoring competitor performance, therefore gaining insights in order to ensure that benchmarking is in alignment with the organizational objectives of management.

In order to combine data from different sources and derive benefit from it, the company must present advanced analytical systems and talented human capital in order to convert data into knowledge. The combination of internal and external data represents a rare and complex asset to imitate: while external data derive from online user interactions, which are within everyone's reach and simple to analyse, internal data derive from organizational processes, which are impossible to be acquired from other sources. Therefore, companies should understand that the value related to the information produced possesses a great potential to creating data-centred business models [15, 27, 16, 28–31].

Figure 3.12 depicts indirect data monetization with reference to product endless improvement. The below diagram presents few modifications essential to describing the transition from a digital-operating model to a data-driven business model. Digital-first products are introduced within the data-driven business model, whilst consumers become connected, due to the consequent product–user interactions. Data enable organizations to connect with users through digital-first objects. Digital-first goods imply constant connection with consumers. As organizations increase product performance, companies eventually realize product endless improvement, aimed at continuously enhancing product performance, driving revenues, as well as increasing production volume.

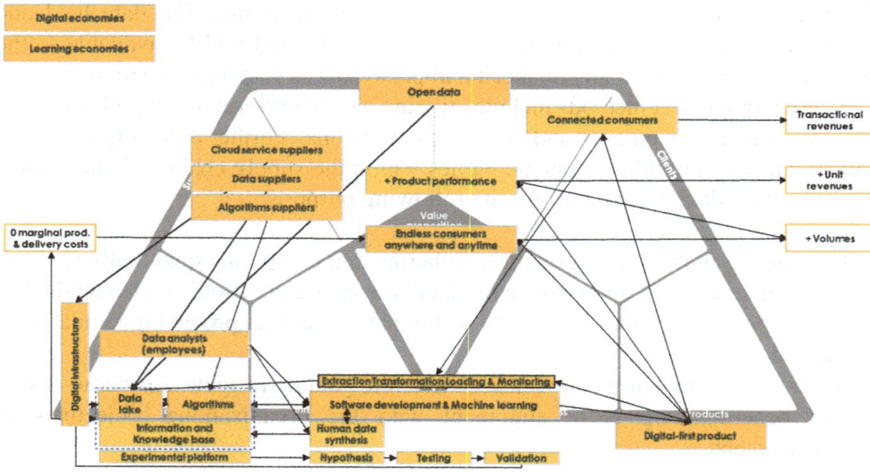

Fig. 3.12 The data-driven business model: indirect data monetization—product endless improvement

3.3.1 Data-Driven Business Model: Indirect Data Monetization—Product Endless Improvement

Apple Inc.

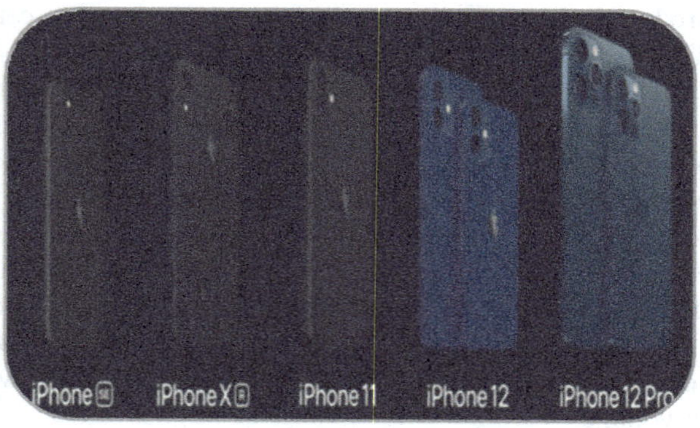

Apple Inc. is an American multinational technology company designing, developing, and distributing consumer electronics, computer software, and online services. Apple launches multiple products and services throughout the year. Typically, Apple releases new iPhone versions yearly characterized by slight design and function improvements, thereby pursuing product endless improvement. For instance, the latest iPhone (iPhone 12) includes the following technological improvements: a better display, improved battery life, enhanced speed, and an innovative dual-camera system, as opposed to iPhone 11.

3.3.2 Indirect Data Monetization: Customization

The advent of big data, artificial intelligence (AI), as well as predictive analytics, and the consequent introduction of further disruptive technologies within organizations have drastically changed customer relationships. Digital-first products require a significant shift within the producer–consumer relationship, currently revolving exclusively around the user experience and favouring the exchange of information during product utilization. The steady interactions occurring between consumers and businesses contribute to generating large amounts of data, gathered and distributed to the central system, thus allowing customer analysis to be conducted effectively. Data and digital technologies constitute the critical drivers fostering a customer-centric culture within companies, as organizations are utilizing customer triggers in order to 'optimize the timing, content, offer, and design of every customer experience' [32]. Organizations conduct a dialogue with individual consumers for the purpose of helping users articulate their needs in order to determine the precise offering fulfilling those requirements, for the purpose of producing customized products for demanding consumers. Furthermore, consumer–product interactions provide organizations with a deeper understanding regarding product utilization, thus converting product usage insights into actionable results, including increased sales as well as enhanced customer satisfaction. Customer loyalty is the consequence of a positive consumer experience as well as a fruitful relationship between consumers and producers.

Figure 3.13 portrays the concepts illustrated above within the business model canvas.

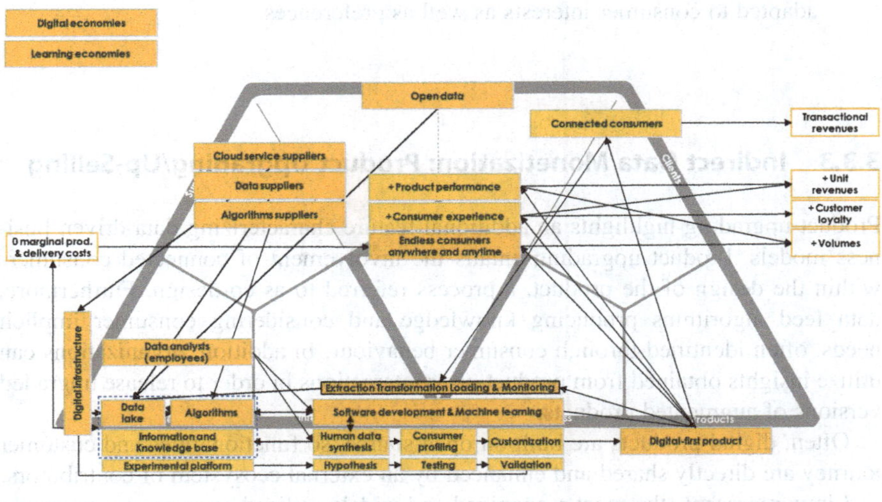

Fig. 3.13 The data-driven business model: indirect data monetization–customization

3.3.2 Data-Driven Business Model: Indirect Data Monetization—Customization
Netflix, Inc.

Netflix, Inc. has been a data-driven company ever since its inception. The core role of analytics is to support business decisions and to gain customer insights. Subsequently, organizations can optimize their digital marketing strategy and deliver a superior product. Data analytics offer businesses the quantitative data crucial to formulating informed decisions and provide a unique and personalized consumer experience. Netflix currently presents millions of worldwide streaming subscribers. Its extensive user base enables the organization to gather an enormous amount of data, to measure and understand engagement, as well as to fulfil consumer requests. Through the process of customer experience management, contents available on Netflix are adapted to consumer interests as well as preferences.

3.3.3 Indirect Data Monetization: Product Upgrading/Up-Selling

Product upgrading highlights an additional feature characterizing data-driven business models. Product upgrading entails the involvement of connected consumers within the design of the product, a process referred to as co-design. Furthermore, data feed algorithms producing knowledge and considering consumer implicit needs, often identified through consumer behaviour. In addition, organizations can utilize insights obtained from product–user interactions in order to release upgraded versions of augmented products.

Often, digital products are built on open source, so functionalities and customer journey are directly shared and enhanced by an external ecosystem of contributors.

Linux represents the most recognized and widely utilized open-source operating system. As an operating system, Linux constitutes software lying underneath all the other software on a computer, receiving requests from computer programs and relaying instructions to the computer's hardware. WordPress represents a free and

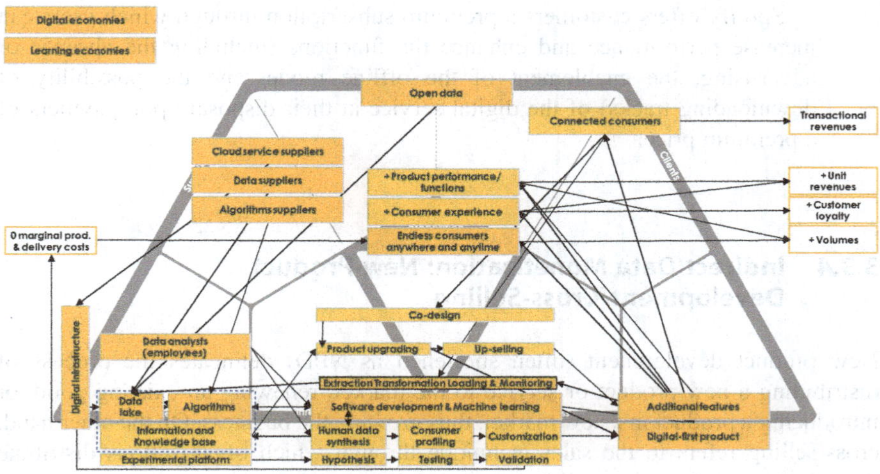

Fig. 3.14 The data-driven business model: indirect data monetization – product upgrading/up-selling

open-source content management system. WordPress stores content and enables users to create and publish webpages, simply requiring a domain and a hosting service.

Product up-selling refers to the process through which additional features are added to a determined digital-first good or service, in order to sell and distribute an enhanced version of a determined product or service, including additional features or enhanced functions relating to the connected product. Up-selling describes a sales technique utilized to persuade a customer to purchase an upgraded or premium version of a determined digital-first product or service.

Figure 3.14 illustrates the process of product up-selling within data-driven organizations.

▶ 3.3.3 Data-Driven Business Model: Indirect Data Monetization Product Upgrading/Upselling Spotify

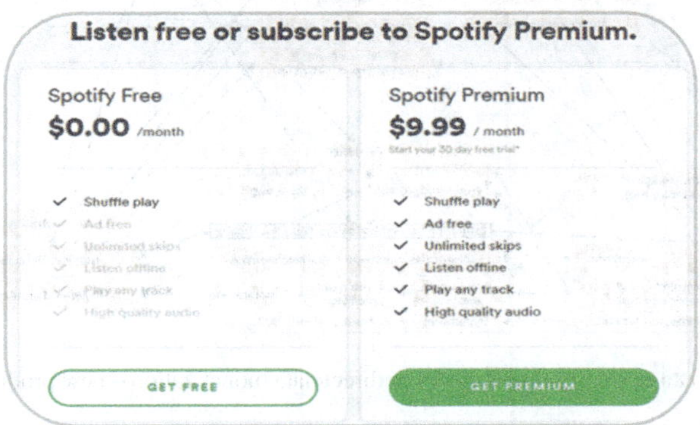

Spotify offers customers a premium subscription through which users can increase performance and enhance the functions (including the absence of advertising, the enablement of the offline mode, and the possibility of downloading tracks) of the digital service at their disposal upon payment of a premium price.

3.3.4 Indirect Data Monetization: New Product Development/Cross-Selling

New product development (often shortened as NPD) delineates the process of distributing a new product or service to the market, renewing an existing good, or introducing a product in a new market, thus growing the business. On the other hand, cross-selling refers to the sales technique through which organizations distribute related or complementary digital-first goods as well as services to customers. Currently, organizations focus on product bundling in order to cross-sell or up-sell. Nonetheless, distributing additional goods to customers implies that consumers gain trust in the company, enabling organizations to invest in building and maintaining loyalty amongst their customers. Businesses successfully achieve economies of scope and gain efficiencies through the production of a wider variety of digital-first products. Efficiencies involve lower average costs as well as increased revenue, as companies increase sales in new, related markets.

Figure 3.15 illustrates the process of product up-selling within data-driven organizations.

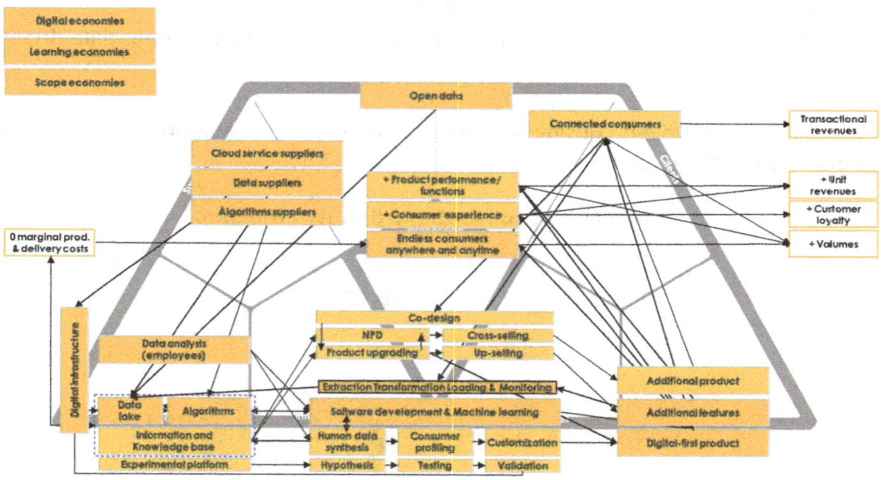

Fig. 3.15 The data-driven business model: indirect data monetization – new product development/ cross-selling

3.3.4 Data-Driven Business Model: Indirect Data Monetization—New Product Development/Cross-Selling

Dyson Limited

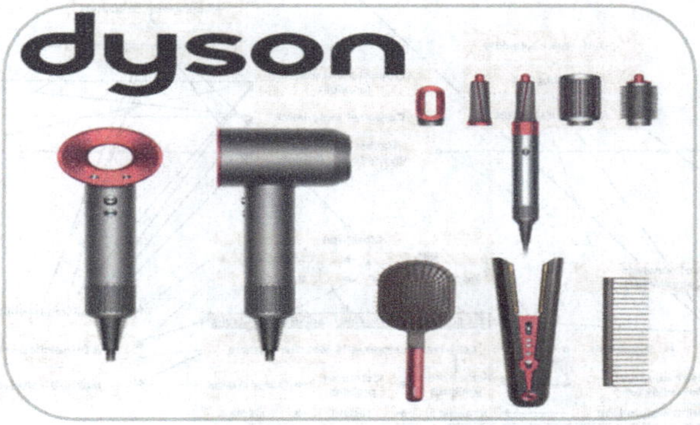

Dyson Limited is a British technology company established in 1991, designing and manufacturing household appliances such as vacuum cleaners, air purifiers, hand dryers, bladeless fans, heaters, hair dryers, and lights. Currently, the company is involved in multiple projects across various industries, as Dyson Limited is planning to design and develop electric vehicles within 2021. The organization's most recognized products include the Dyson Supersonic hair dryer and the Dyson Airwrap, as well as the Dyson Corrale hair straightener. Dyson developed the aforementioned products (including a paddle brush and a detangling comb) in an attempt to offer customers the opportunity to improve their own experience through the purchase of both hair-styling tools, thus pursuing cross-selling opportunities.

3.3.5 Indirect Data Monetization: New Market Space Creation

Organizations developing additional products distribute goods to existing consumers, therefore exploiting cross-selling opportunities or targeting and connecting with prospective customers, potentially generating additional sales leads and increasing revenue from existing clientele. Market segmentation describes the process of dividing existing as well as potential customers into clearly defined groups, establishing distinct sets of criteria associated with each group. The goal consists of grouping together customers with similar needs, assigning a value based on business goals to each category, thus facilitating the organization's ability to identify the most valuable consumers for the business and effectively sell to high-value customers. Companies utilize customer segmentation in order to optimize their ability to market to a wide variety of customers. Furthermore, creating target

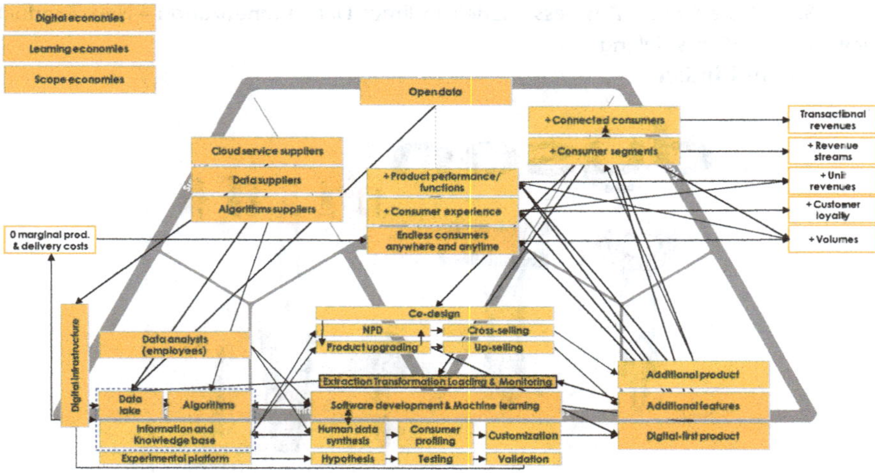

Fig. 3.16 The data-driven business model: indirect data monetization – new market space creation

customer segments and prioritizing sales leads, as well as introducing additional revenue streams represent prerequisites of organizational success.

Figure 3.16 illustrates the process of new market space creation within digital organizations.

3.3.5 Data-Driven Business Model: Indirect Data Monetization

New Market Space Creation.

Amazon.com, Inc.

Amazon distributes downloads and streaming of video, music and audiobooks through its Prime Video and Amazon Music. Amazon also includes a publishing arm, Amazon Publishing, a film and television studio, Amazon Studios, as well as a cloud computing subsidiary, Amazon Web

Services. In addition, Amazon produces consumer electronics including Kindle e-readers. Once consumer preferences are collected, Amazon is able to distribute goods as well as services in line with the needs of its customer base.

3.3.6 Direct Data Monetization

Direct data monetization refers to the process of granting a third-party access to organizational data in order to generate a direct revenue stream. In addition, direct data monetization presents multiple advantages, including expanding customer relationships. External or direct methods include data sharing in order to gain beneficial terms or conditions from business partners, information bartering, selling data outright, as well as offering information products or services. Direct monetization complements indirect monetization. The major capabilities necessary in order to maximize the value of digital data include the following: the implementation of advanced analytical systems to combine data from different sources and to ensure their effective utilization as well as talented human capital in order to translate data into insights.

Figure 3.17 pictures direct data monetization within the business model canvas.

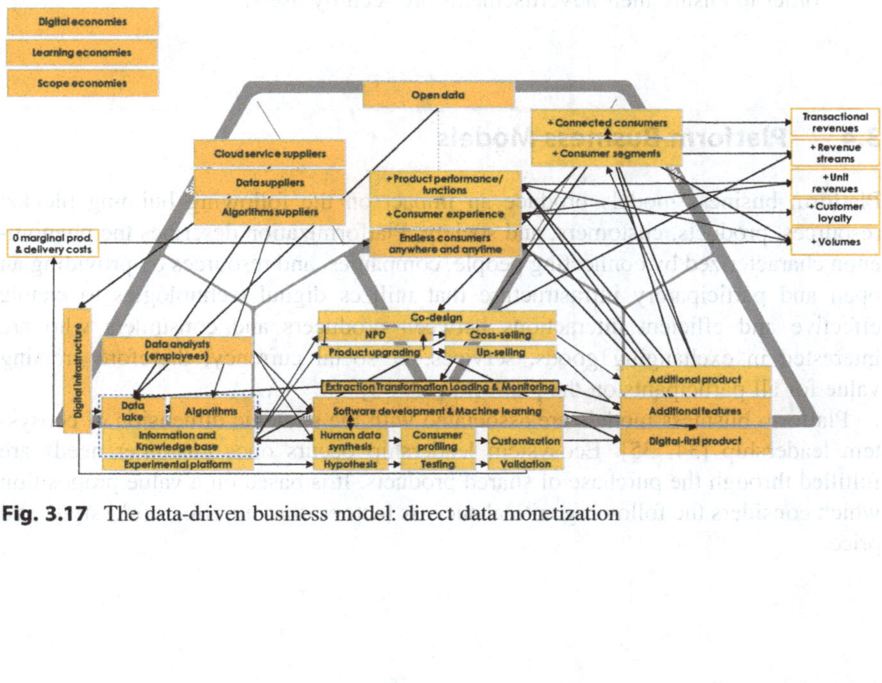

Fig. 3.17 The data-driven business model: direct data monetization

3.3.6 Data-Driven Business Model: Direct Data Monetization Facebook, Inc.

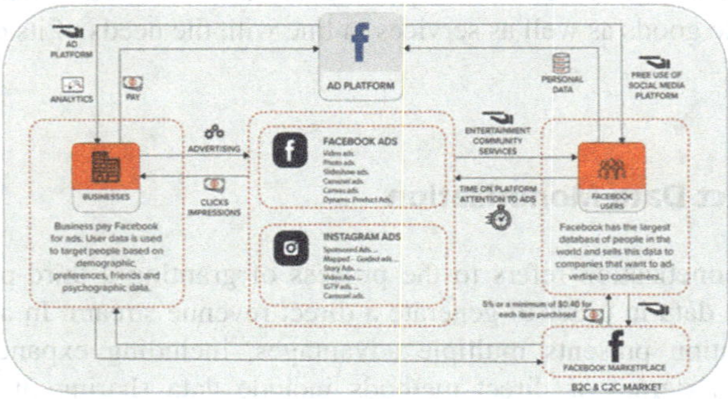

The collection of data constitutes an asset which, in certain contexts, is directly monetized. Facebook itself profits from personal as well as online data. Although Facebook offers a free service, the social network captures data and user engagement in a manner which can be monetized. In particular, Facebook tracks its users across the web utilizing tracking cookies. Once users are logged into Facebook and simultaneously browse additional websites, the social media platform is able to track the internet sites which community members previously visited. Utilizing vast amounts of personal data enables the organization to tailor advertisements in order to suit an individual's situation. Targeted advertising represents a form of online advertising delivering promotional messages to users according to their specific traits, interests, as well as preferences. Furthermore, targeted advertising is considered to be beneficial to advertisers, who are willing to pay a premium in order to ensure their advertisements are seen by users.

3.4 Platform Business Models

Platform business models produce an impact on the following building blocks: resources, products, customers, and society. Platformization describes the phenomenon characterized by connecting people, companies, and resources by providing an open and participatory infrastructure that utilizes digital technologies to enable effective and efficient interactions between producers and consumers who are interested in exchanging goods, services, or social currency, therefore creating value for all participants on the platform [13, 33].

Platform business models are associated with the strategic dimension of ecosystem leadership [34, 35]. Ecosystem leadership occurs once customer needs are fulfilled through the purchase of shared products. It is based on a value proposition which considers the following critical success factors: sharing, life (work) style, and price.

		JAN.	NOV.	GROWTH			JAN.	NOV.	GROWTH
1.	Zoom	11%	26%	15.1	11.	Roundup	12%	17%	4.6
2.	Peacock	4%	16%	11.5	12.	Clorox	62%	66%	4.5
3.	TikTok	10%	17%	6.9	13.	AstraZeneca	6%	11%	4.5
4.	Instacart	10%	17%	6.3	14.	Cash App	18%	22%	4.5
5.	DoorDash	22%	28%	6.1	15.	Twitch	8%	12%	4.3
6.	HBO Max	16%	22%	6.1	16.	Amazon Fresh	18%	22%	4
7.	WhatsApp	14%	19%	5.1	17.	Venmo	17%	21%	3.7
8.	Microsoft Teams	18%	23%	4.8	18.	Tubi	12%	15%	3.6
9.	T-Mobile	22%	27%	4.8	19.	Pluto	13%	17%	3.6
10.	Pfizer	14%	18%	4.6	20.	BBC	19%	23%	3.6

Fig. 3.18 The 20 fastest growing companies of 2020

Currently, traditional businesses are converting into successful digital businesses, contributing to changing consumer markets as well as renovating industries. Most corporations implement the aforementioned model for the purpose of growing exponentially as well as of capturing significant market share from established companies. Presently, platform-based business models are emerging in multiple sectors, including retail, education, and transport. Platform business models require the digitization of three major elements, including consumer behaviour, identity and trust, and location. The digitization of consumer behaviour has allowed companies to collect and utilize data for the purpose of influencing consumer purchasing decisions. In addition, digital identification systems have contributed to maintaining trust in order to enable transactions to be conducted seamlessly and securely. Moreover, the introduction of smartphones and Internet of Things (IoT) devices has contributed to locating and connecting individuals as well as digital items. Digitalization allows conversion of physical items into digital objects and to receive real-time location data of multiple digital-first products. Organizations are currently implementing software systems mediating efficient social as well as business interactions, thus adopting a plug-and-play business model allowing connected users and objects to plug in and orchestrate consumers towards value-creating interactions.

Fifteen of the twenty most valuable companies globally have implemented a digital business model. In particular, 10 of the organizations reported in Fig. 3.18 are now based on a platform business model, thus creating digital communities as well as marketplaces, allowing distinct groups to interact and transact. Most of the businesses included within the list are connected to pandemic-related behaviour, ranging from at-home entertainment to pharmaceutical companies. The figure compares the percentage growth companies experienced during November 2020 as opposed to January 2020. The growth index was computed by determining the share of consumers who considered purchasing goods and services from the companies illustrated below during November 2020 minus the share of customers who reported similar answers, two months later [10]. McKinsey & Company hypothesizes that by 2025 more than 30% of global economic activity is going to be mediated by digital platforms, while only a small share of established companies has adopted an effective platform strategy [10].

Platform business models favour as well as enhance value-creating interactions between external producers and consumers. Furthermore, platforms are characterized by an open and participative plug-and-play infrastructure for suppliers and customers to plug into and interact with one another. Platform-based business models comprise four major characteristics: connectivity, intelligence, openness, and scalability. As reported by Rita Gunther McGrath in the paper entitled 'Is your company ready to operate as a market?', the Fourth Industrial Revolution has contributed to the emergence of platform businesses [36]. Airbnb, Uber, and Alibaba represent a few examples of companies leveraging digital platforms. The aforementioned organizations are commonly referred to as inverted firms; inverted firms are achieved by shifting the company's production to ecosystem partners. Moreover, an open system allows for multiple opportunities and orchestration to innovate and solve issues. The shift has contributed to considerably increasing development as well as growth, as platform firms do not present direct costs and their pace of growth is determined by their ability to acquire additional partners. Platform business models offer consumers a comprehensive digital experience rather than a mere good or service. Furthermore, products lose their value almost immediately, as consumer goods rapidly depreciate, whilst the value of platforms increases as the service is utilized. In this regard 'value creation attracts greater value consumption' [37]. The value of the user experience increases along with the rising number of individuals and organizations involved as well as through complementary innovations.

Business ecosystems represent a network of companies, including further individuals and organizations, involved in the delivery of particular products or services through both competition and cooperation. Participation within a business ecosystem provides mechanisms in order to leverage technology, to achieve excellence in research, innovation, and business competence, as well as to compete effectively against additional enterprises. Furthermore, ecosystems allow acceleration of the learning process to effectively collaborate and share insights, knowledge, and expertise. Ecosystems of users represent an innovative source of supply and value creation as well as of competitive advantage and market dominance, thus demonstrating the importance of the digital shift to technologies in a globally connected world.

Amongst the major elements characterizing platform firms, it is worth mentioning *coring* and *tipping*.

The former (*copping*) represents a set of activities which companies utilize in order to identify or design technologies and products as well as to render them fundamental to a particular platform or a market. Platforms are available to partners as well as customers who have access to specific technologies and who contribute to developing additional utilization modes, thus contributing to the overall growth of the digital structure. Encouraging further businesses to join the ecosystem also represents an important aspect related to the enhancement of platform firms. Google represents a significant example of coring, as the enterprise has established itself as the leader in search technologies as well as a foundational structure in Internet navigation. Google started out solving a major issue: finding all sorts of information within the Internet, later, it distributed its technology to website developers and users, facilitating further development opportunities originating from it.

The latter (*tipping*) refers to the set of strategic activities which companies execute to emerge as dominant players within the market. Actions range from the formulation of sales strategies to marketing, product development, and alliance-building. Successful tipping occurs once organizations can develop unique features that appear difficult to imitate, therefore attracting users. Furthermore, tipping can be realized once competitors and users join together in a coalition, consisting of a defence mechanism, to prevent and restrict competition. Linux represents a notable example of successful tipping within the IT sector. It has evolved through a community of open-source programmers and users, becoming the fastest growing operating system in the world [38].

Current businesses achieve scale by leveraging interactions within the ecosystem; specifically, 'platform businesses scale through network effects' [35]. Network effects contribute to enhancing value creation, as value is generated and exchanged through user interactions. Network effects represent the primary source of value creation and competitive advantage in a platform business and describe the incremental benefit gained by an existing user for each additional participant joining the network. Furthermore, network effects can be distinguished into two categories: positive and negative. Positive network effects are determined by the ability of a wide platform community to produce significant value for each participant, whilst negative network effects arise due to the possibility that a poorly managed platform community might contribute to reducing the value produced for each user. The effects reported above allow to stimulate network growth and often encourage organizations to spend money for the purpose of attracting participants to one side of the market, as once platform businesses have encouraged consumers to join the platform, producers are going to connect to the platform as well. Moreover, two-sided networks present four kinds of network effects, including positive or negative same-side effects as well as positive or negative cross-side effects. Same-side effects, also referred to as direct network effects, comprise the positive benefits received by either producers or consumers, once the number of users of the same category increases. On the other hand, cross-side effects, also referred to as indirect network effects, emerge as either producers or consumers gain or lose based on the number of users on the opposite side of the platform.

Uber represents a notable example of network effects. The value of the service offered increases as the number of individuals utilizing the service grows. Nevertheless, in the event that the service attracts a greater number of drivers compared to the number of riders, driver downtimes are expected to rise, whilst in case Uber attracts an increasing number of riders as opposed to drivers, rider waiting periods are expected to increase as well (as scaling a network requires that both sides constituting the digital ecosystem grow proportionally). Organizations operating in

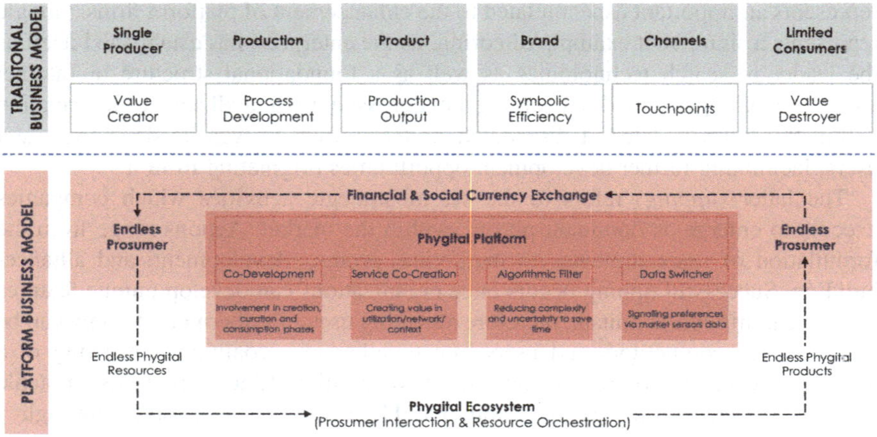

Fig. 3.19 The platform business model

a two-sided market must manage all four network effects. In addition, platform businesses are required to reinforce positive network effects, thus creating and strengthening positive feedback loops.

Figure 3.19 serves to report the major differences between traditional pipeline models and platform business models. The former is focused on internal processes, as firms design and develop products and promote brand engagement. Subsequently, retailers act as middlemen between wholesalers and the end customers, thus connecting businesses with consumers. Ultimately, consumers purchase products or services. By contrast, platform businesses are centred around a core interaction involving two participative roles, namely producers and consumers, in the business ecosystem. Technological advancements allow to connect independent groups of users to each other, as interactions are crucial for the growth and the value creation of the platform to its participants. Nonetheless, platforms cannot control quality as traditional pipeline models used to. Platforms cede control to the ecosystem, therefore failing to control the value creation process. Platform-based businesses must implement strict quality control mechanisms, as most platforms require initial screening of producers for the purpose of ensuring a minimum quality threshold. Algorithms and machine-learning techniques serve to replace traditional gatekeepers as well as to aggregate and match demand and supply fairly and efficiently, as platform consumers are served in a highly customized manner by leveraging data and analytics across the entire organization. Furthermore, platforms utilize data for the purpose of automating decision-making processes as well as of digitizing processes. In addition, ecosystem interactions are orchestrated utilizing data. Moreover, social feedback represents a major source of user incentivization on platforms.

Platform-based business models involve three significant shifts affecting the value creation process.

- **Shift in markets: from consumers to producers.** Following the traditional view of the market, the pipe would distribute products and services to customers, thus locating consumers at the end of the pipe. There used to be a direct relationship with the business, as organizations would build goods which consumers would request and customers would pay in exchange for determined products or services, usually with money as well as attention and engagement. By contrast, platform businesses do not create the end value, instead organizations only enable value creation. Participants act as producers as well as consumers. Platforms must encourage suppliers to engage consistently, otherwise the business would be unlikely to become successful at creating value.
- **Shift in competitive advantage: from resources to ecosystems.** Platform business models do not own the means of production, alternatively platforms create the conditions for connection, thus leveraging data in order to orchestrate physical as well as digital resources across the ecosystem. Platforms successfully orchestrate value-exchanging interactions within the ecosystem utilizing data regarding multiple ecosystem participants. Ecosystems represent the key enablers of value creation on platforms as well as an innovative source of competitive advantage.
- **Shift in value creation: from processes to interactions.** Platforms facilitate interactions amongst participants, thus determining value creation and exchange. Value is created and scaled through interactions, orchestrating users and resources within the ecosystem [35].

'Platforms must perform three key functions in order to encourage a high volume of valuable core interactions' [34], commonly summarized as: attract attention, build a reputation, and enable monetization.

- **Attract attention.** Platforms must attract participants to the platform, therefore enabling interactions amongst users. Attracting consumers to platforms presents particular challenges which traditional organizations do not face. Platforms must learn to overcome the chicken-or-egg problem, as users are not going to join the platform unless the platform generates value. Nonetheless, a platform is not going to generate value unless participants utilize the service. Furthermore, platforms must maintain user interests and the primary tool encouraging participants to utilize the service consists of the feedback loop. Single-user feedback loops represent a powerful tool in order to increase activity and value creation as well as to enhance network effects, since the more participants utilize the platform, the more information the platform collects about users and the more accurate its recommendations become. Platform founders are currently searching for solutions to the dilemma referred to as the chicken-or-egg problem.
  Amazon never suffered from the chicken-or-egg problem as a successful online retailer it operated an effective pipeline business, utilizing online product listings in order to attract consumers, thus building its platform business on the foundation of an existing pipeline business. Amazon Marketplace converted itself into a platform-based business once it opened its business to external producers, thereby

enabling merchants to sell products to consumers, with Amazon retaining 10–15% of each transaction.

- **Build a reputation.** Platforms must facilitate interactions, offering users the tools and instructions in order to enable producers and consumers to connect and encourage valuable exchanges. Facilitating interactions often involves lowering barriers to utilization, thus encouraging and expanding participation on the platform. In order to create and maintain a good reputation, platform businesses ensure consumers can access their personal information, thus requiring businesses to implement a clear privacy and data management policy. Moreover, digital businesses must guarantee the reliability of the digital infrastructure and the security of transactions, therefore adopting data security measures. In addition, platforms must verify the identity of participants within the ecosystem and define mandatory requirements in order to operate. Furthermore, the orchestrator must guarantee process efficiency and a smooth user experience on the platform. Platforms nurture trust once the business encourages its customers to inform each other about the service offerings, therefore stimulating exchanges of information as well as experience.
- **Enable monetization.** Platforms must connect producers and consumers effectively by utilizing information regarding each for the purpose of connecting participants and building mutually rewarding and beneficial relationships. Network effects render platforms attractive through the creation of self-reinforcing feedback loops, contributing to growing the user base. Nonetheless, charges levied on users are likely to discourage participants from joining the platform. Charging for access may lead individuals to avoid the platform, while charging for utilization may inhibit frequent participation; charging for production reduces value creation and charging for consumption contributes to reducing consumption. Platform companies are engaged in designing and building technology, inviting participants to join the platform in order to monetize the business by charging for the value that the platform technology creates for users. A smart monetization strategy considers four distinct forms of value for the purpose of determining the sources of excess value exploited by the platform, without inhibiting the continuous growth of network effects. Value falls into four broad categories:
  - For consumers: access to value creation. Charging a transaction fee represents a mode of monetizing the value generated by the platform, without hindering the growth of network effects. Consumers and producers are going to be charged in the event an actual transaction occurs, thus not discouraging users to join the platform.
  - For producers as well as third-party providers: access to a community or market. A further option is represented by the possibility to monetize a platform by charging producers for access to a community of users, who have joined the platform for unrelated reasons. Charging third-party producers for community access is effective in case the added contents, such as job opportunities on LinkedIn, enhance the value of the platform to its participants.

- For consumers and producers: access to tools and services facilitating interaction. Charging producers for enhanced access to consumers, enabling suppliers to be noticed on a two-sided platform, despite the abundance of producers and the resulting intense competition in order to attract consumer attention. Platforms charging producers fees for attractive presentations, or interactions with particularly valuable participants, are utilizing enhanced access as a monetization technique.
- For consumers and producers: access to curation mechanisms enhancing the quality of interactions. Once the amount of content on a platform becomes greater, consumers might struggle to find the high-quality content users are searching for, thus reducing the platform's value for participants. In this event, consumers may be willing to pay for access to guaranteed quality, namely enhanced curation [34].

'Across all platforms, the following three distinct layers emerge repeatedly: network- marketplace-community, infrastructure and data' [35].

- **Network-Marketplace-Community Layer.** The first layer of a platform consists of the network or community layer, comprising the participants on the platform and their valuable relationships. Social media platforms require users to connect with each other explicitly, whilst marketplaces may not require to establish explicit connections, alternatively marketplaces may regularly match consumers and producers for the purpose of enabling desirable interactions. Multiple users benefit implicitly from participation within the digital community, although not connecting with further participants explicitly. The external network of producers creates value within the network layer. Nonetheless, in order to successfully enable value creation, platforms require a second layer, referred to as the infrastructure layer.
- **Infrastructure Layer.** The infrastructure layer encapsulates the tools, services, as well as the directions enabling the plug-and-play nature of a platform business. Nonetheless, the infrastructure layer presents little value, unless participants create value on the platform. External contributors build on top of the infrastructure. Furthermore, the infrastructure layer provides the digital infrastructure on top of which value can be generated. Platforms constitute open systems, allowing a multitude of external participants to create value. Large-scale value creation as well as increasing participation encourage unrestricted production, usually leading to abundance, thus resulting in higher search efforts for consumers. In order to overcome the issue, platforms resort to a third layer, referred to as the data layer.
- **Data Layer.** The data layer constitutes the final platform layer. Digital ecosystems utilize data, as data collection contributes to supporting an adequate matching of supply and demand. In this regard, filters serve for the aforementioned purpose. The platform's ability to filter relevant content significantly improves with increased utilization, as interactions amongst participants occur, the ecosystem captures signals regarding the contents that users consider desirable, thus leveraging data in order to strengthen filters further. The data layer

promotes relevance, therefore matching suitable content as well as goods or services to consumer preferences [35].

The platform business model is further classified into four distinct categories: data confederation, infrastructure community, social network, and marketplace. The subsequent paragraphs aim at illustrating the classification proposed within the business model canvas.

3.4.1 Data Confederation

Data constitute the principal asset characterizing digital ecosystems. Platforms gather data on the producer side for the purpose of delivering valuable information and knowledge to consumers. Moreover, the platform business model ecosystem enables interactions around a digital infrastructure. Platforms enable unlimited access to all users willing to join the ecosystem, thus contributing to guaranteeing access to an infinite number of participants, continuously engaging in fruitful interactions. Furthermore, the data confederation business model is characterized by the introduction of prosumers. The term represents a portmanteau of the following words: producer and consumer. The expressions of prosumer and prosumption were coined by Alvin Toffler, an American futurist, in 1980. Following his perspective, marketplaces became increasingly saturated as mass production of standardized goods began to satisfy basic consumer demands. Nonetheless, in order to continue to grow profit, businesses initiated a process of mass customization, namely the mass production of highly customized products. Consumers are required to participate within the production process, thus defining particular design requirements for the purpose of achieving a greater degree of customization. Technological advancements as well as the increase in user participation blur the boundaries between production as well as consumption activities, consequently consumers become prosumers. Prosumers play crucial roles in co-production, creation, promotion, as well as distribution through interactive conversations and dialogues amongst digital communities. The diffusion of innovation alongside the evolution of social media platforms have contributed to enhancing the process of prosumer engagement. In addition, the user base characterizing digital ecosystems contributes to creating a digital community, therefore producing significant networking effects. Indirect relationships arise as users contribute data (often unconsciously) in order to provide relevant information to further participants within the ecosystem for the purpose of enhancing user experiences. Platform-based businesses, presenting a large user base, collect and evaluate a considerable amount of data in order to significantly improve the prosumer experience [1].

Figure 3.20 depicts the concepts illustrated above.

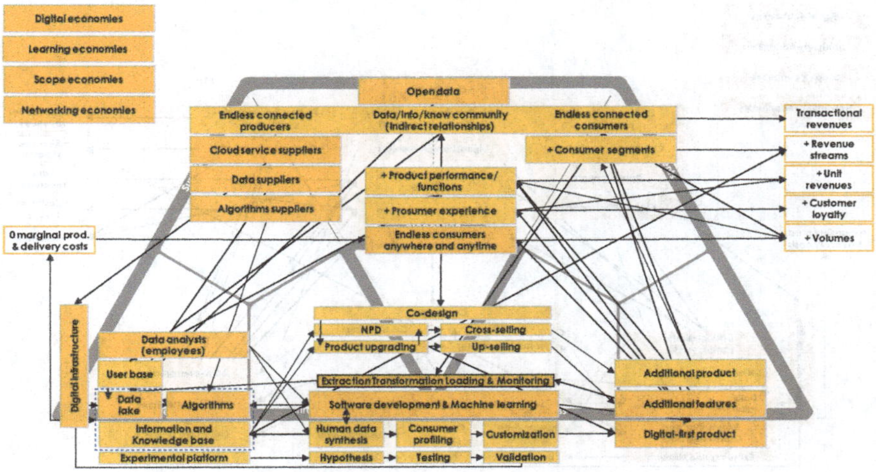

Fig. 3.20 The platform business model: data confederation

3.4.1 Platform Business Model: Data Confederation TripAdvisor, Inc.

TripAdvisor has created an entirely new industry based on certifying the quality of product and service providers, while in the process driving few traditional industry certifiers out of business. Furthermore, participants within the digital ecosystem contribute large quantities of data for the purpose of evaluating prosumer experiences. In this regard, value created during individual interactions persists to create cumulative value. Platforms determine producers' or consumers' reputation and quality by aggregating social signals from the community. Furthermore, community members on TripAdvisor pay

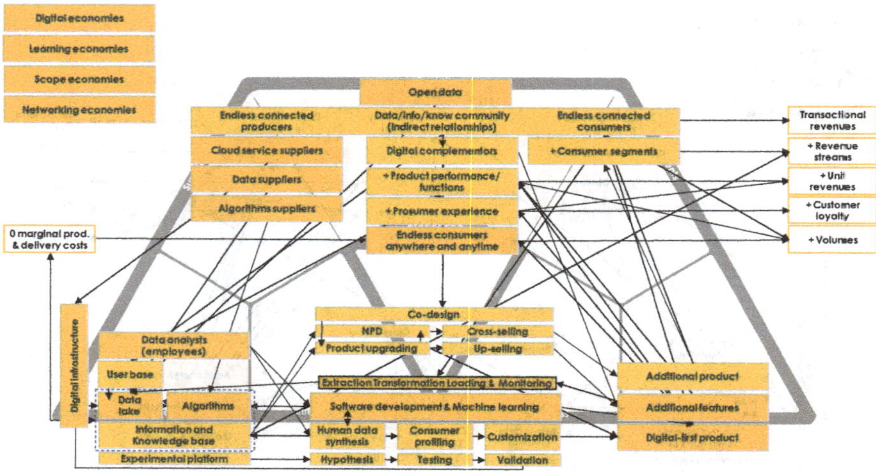

Fig. 3.21 The platform business model: digital complementors

by enhancing the reputation of producers whose work they appreciate. Thus, fame, influence, reputation, and additional intangible forms of value constitute a particular kind of 'currency' on a platform.

Digital complementors represent a feature characterizing the platform business model. Developing digital ecosystems requires to consider the modalities through which digital organizations can leverage goods and services offered within the ecosystem as well as services provided by further technology companies for the purpose of improving user experiences (Fig. 3.21).

3.4.2 Infrastructure Community

Platforms provide an open and participative infrastructure that users plug in to, for the purpose of enabling value-creating interactions and of setting conditions governing relationships amongst participants as well as of establishing rules rendering exchanges simple and mutually rewarding. Removing barriers to production as well as consumption enables platform firms to scale interactions. Plug-and-play infrastructures enable members of the technology community to actively engage with contributors in the digital ecosystem.

Furthermore, platform providers render accessible technological tools to application developers in order to design applications, therefore participants are involved both in the co-design as well as in the co-development of the platform in order to expand and enhance the service offerings. Nonetheless, users can either develop software, thus becoming producers, or utilize the software generated by members of the digital community and begin developing further software components from

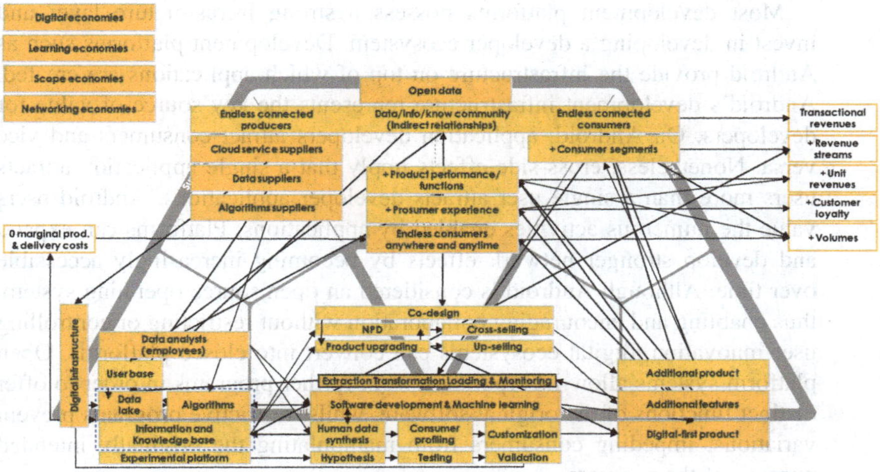

Fig. 3.22 The platform business model: infrastructure community

existing software already available on the platform. In this regard, application programming interfaces (often shortened as APIs) represent sets of shared protocols, enabling developers access to a proprietary software application for the purpose of creating an innovative application software. In addition, application programming interfaces (APIs) render the features of multiple applications available in order to simplify the design and development of additional services in response to corporate needs. Organizations often render their application programming interfaces (APIs) public in order to allow suppliers to develop further applications, exploiting the functionalities already embedded within the API. Exposing APIs to the public enables digital organizations to facilitate openness to innovation as well as to increase efficiency due to collaboration with external contributors.

Figure 3.22 displays the concepts described above.

► **3.4.2 Platform Business Model: Infrastructure Community**

Android

developers

Most development platforms possess a strong infrastructure layer and invest in developing a developer ecosystem. Development platforms such as Android provide the infrastructure on top of which applications are created. Android's development infrastructure represents the key source of value for developers. On Android, application developers attract consumers and vice versa. Nonetheless, cross-side effects imply that a single application attracts users more than a single user attracts developer applications. Android users value the numerous activities enabled by applications. Platforms can expand and develop stronger network effects by becoming increasingly accessible over time. Although Android is considered an open source operating system, thus enabling and encouraging collaboration without restricting or controlling user innovation, digital ecosystems can convert into closed platforms. Open platform systems allow incorporation from further programs in order to offer distinct functions of the original software, whilst restrictive programs prevent variations, impeding consumers from manipulating the originally intended purpose of the program.

3.4.3 Social Network

Networking platforms aim at creating explicit connections amongst members of a digital ecosystem in order to generate mutually beneficial opportunities. Hence, the role of platforms consists of enabling interactions amongst participants, generally on a digital infrastructure. Interactions also enable the creation of lasting as well as persistent value. Digital ecosystems are introduced, and direct personal relationships emerge as the individuals with whom members occasionally interact on the platform are typically friends. On social media platforms, producers provide endless digital resources, including contents, thus generating valuable interactions. Nonetheless, the introduction of digital resources implies the adoption of particular rules (referred to as information accessing and controlling) in order to determine whether producers fulfil certain criteria and to implement advanced mechanisms for quality control, while encouraging active participation by an ecosystem of producers. Furthermore, curation typically comprises screening and feedback at critical points of access to the platform. Screening determines the individuals to let in, whilst instant feedback encourages desirable behaviour on the part of the users who have been granted access. Often, users curate the platform, utilizing software tools enabling members to gather and aggregate feedback and apply it to curation decisions. In addition, information filtering serves to determine the suitability of digital contents for determined users and to acquire large amounts of data regarding consumers for the purpose of delivering value to customers. Social currency represents a subjective and intangible form of value as well as the influence of members on social networks. In particular, social currency (measured as the economic value of a relationship) includes favourites and shares as well as reputation individuals build as a result of valuable interactions. Furthermore, the platform's overarching purpose includes: 'to

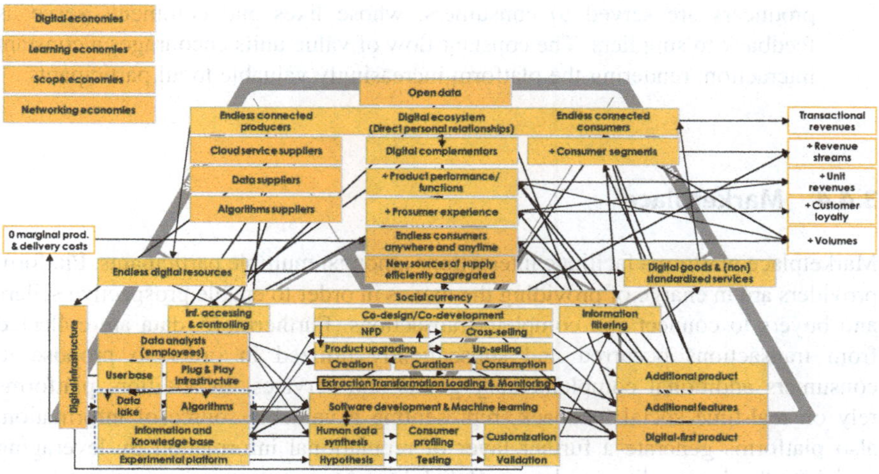

Fig. 3.23 The platform business model: social network

consummate matches among users and facilitate the exchange of goods, services or social currency, thereby enabling value creation for all participants' [34].

Figure 3.23 illustrates the aforementioned concepts within the business model canvas.

3.4.3 Platform Business Model: Social Network Facebook, Inc.

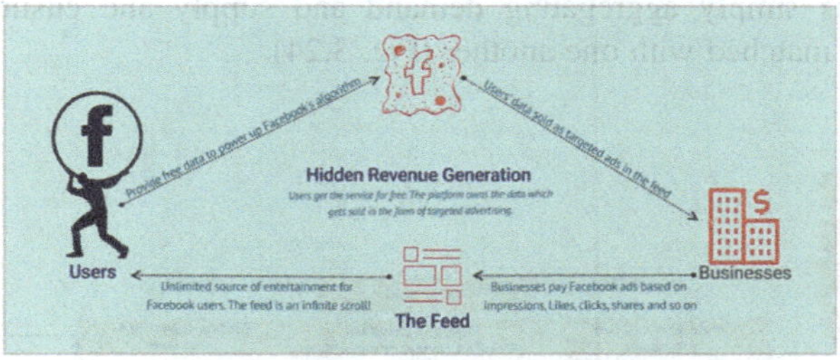

Amongst the most popular social media platforms of the present generation, Facebook represents a disruptive as well as a multisided platform, creating no content. Facebook is considered valuable due to the digital communities participating within the social media platform. On Facebook, photos as well as posts containing personal information are exchanged amongst users. Platforms capturing mainly attention can monetize the business collecting payments from third parties, considering the attention valuable. On Facebook, advertisers are willing to pay the digital platform for users attracted by posts related to a particular topic. Furthermore, the social media platform's news feed represents a classic multiuser feedback loop, as status updates from producers are served to consumers, whose likes and comments serve as feedback to suppliers. The constant flow of value units encourages a constant interaction, rendering the platform increasingly valuable to all participants.

3.4.4 Marketplace

Marketplace platforms facilitate interactions amongst multiple participants. Platform providers are in charge of providing the means in order to enable prospective sellers and buyers to connect and complete transactions. Furthermore, data are collected from transactions occurred and subsequently utilized in order to propose to consumers additional complementary goods and services. In addition, platforms rely on real-time social feedback, representing a valuable source of information, also platforms generate a further layer of reputational information by leveraging social feedback regarding producers. Reliability of transactions represents a relevant aspect as well. Transactions involve physical as well as digital resources (referred to as phygital resources), whilst (non) standardized services describe the process through which marketplaces propose regular items to consumers based on data collected, enabling the platform to identify user preferences and requirements, and therefore offering customers goods suiting their needs. Moreover, users purchase goods and services, thus paying a price for the determined product or service acquired. The goal of a platform business is to connect producers and consumers and ensure both participants engage in three distinct forms of exchange: of information, of goods or services, and of currency. Furthermore, platforms create efficiencies in interactions simply aggregating demand and supply and ensuring users are appropriately matched with one another (Fig. 3.24).

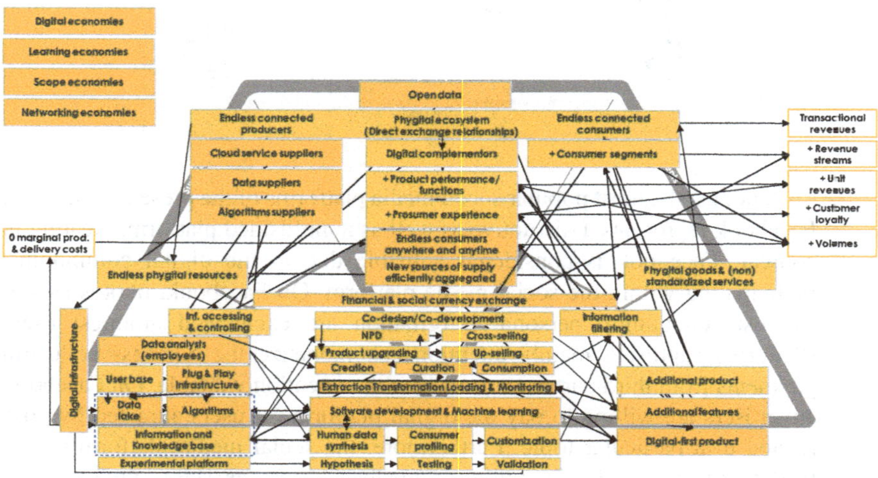

Fig. 3.24 The platform business model: marketplace

3.4.4 Platform Business Model: Marketplace Airbnb, Inc.

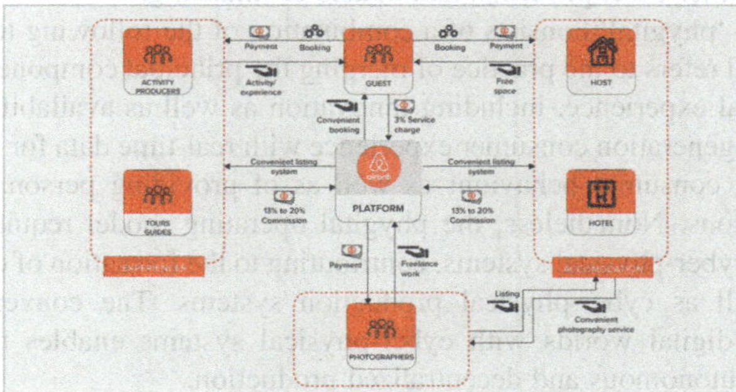

Airbnb, Inc. managed to disrupt and dominate vast traditional industries within a few years. Airbnb constitutes a global provider of accommodation, applying the platform model to the hotel business, as it does not own real estate. The platform creates the conditions enabling individual participants to provide the rooms directly to consumers. In return, Airbnb retains a 9–15% transaction fee for every rental arranged through the platform. Airbnb hosts and guests review each other during interactions, thus creating reputation enabling future transactions. On Airbnb, hosts attract guests and vice versa, therefore attracting two-sided network effects with positive feedback. Typically, the growth of a platform can be facilitated and achieved through the phenomenon referred to as side switching effect, occurring in the event that users pertaining to one side of the platform join the opposite side. In this regard, Airbnb recruits hosts from amongst its guest pool. Amidst the Covid-19 crisis, Airbnb has been able to transform a desperate situation into an opportunity to increase revenues, thus diversifying its initial lodging offer and entering a growing experiential travel sector. Airbnb Experiences represent online as well as interactive activities designed by inspiring local experts and hosted for anyone, from anywhere. Airbnb offers users the possibility to connect with multiple travellers and locals, seeking unique experiences which may not be available through more traditional channels.

3.5 Phygital Operating Model

The phygital operating model originates from the digital operating model and constitutes the basis for the development of the smart factory business model. As previously mentioned, the operating model represents a visualization outlining the modalities through which organizations operate in order to deliver value to their customers or beneficiaries. Operating models represent a subset of a broader concept referred to as the business model. By contrast, the business model defines the plan for the successful operation of the business, thus identifying sources of revenue, the intended customer base, products, and details of financing.

The notion 'phygital' consists of a combination of the following terms: physical and digital and refers to the practice of merging the principal components associated with the digital experience, including innovation as well as availability in order to fulfil the next-generation consumer experience with real-time data for the purpose of understanding consumer behaviour as well as of providing personalized product recommendations. Nonetheless, the phygital operating model requires the implementation of cyber-physical systems, contributing to the formation of cyber-physical plants, as well as cyber-physical production systems. The convergence of the physical and digital worlds with cyber-physical systems enables the innovative paradigm of autonomous and decentralized production.

The Fourth Industrial Revolution is characterized by a more optimal utilization of material resources as well as human resources, in particular, resource optimization is realized as a result of the implementation of cyber-physical systems as well as cyber-physical production systems (often shortened as CPS and CPPS).

Cyber-physical systems (CPS) consist of product systems constituted of physical as well as virtual components. Physical components represent tangible devices equipped with:

- Sensors, enabling cyber-physical systems (CPS) to collect structured as well as non-structured data and to perceive the surrounding environment within which the object circulates as well as to enable cyber-physical systems (CPS) to be controlled (both physically and virtually) by further physical items.
- Transmitters and receivers, facilitating interactions amongst multiple individuals and digital devices.
- Memory, allowing cyber-physical systems (CPS) to store and exchange data previously gathered, thus drastically reducing processing times as well as reaction times.
- Processors, autonomously transforming data into information and knowledge and ultimately decisions, through algorithm design techniques.
- Actuators, enabling cyber-physical systems (CPS) to modify their operation in accordance with the decisions taken and to adapt to the surrounding environment.

Furthermore, the physical component allows the cyber-physical system (CPS) to monitor itself in real-time in terms of effectiveness and efficiency throughout its entire life cycle. The monitoring function enables the components constituting the cyber-physical system (CPS) to:

1. decide on the best operating conditions;
2. flexibly adapt in order to respond to changes in consumer demand; and
3. learn autonomously without requiring the supervision of human operators [39, 40].

Therefore, smart product systems can be defined as connected, adaptive, customizable, and proactive, as cyber-physical systems (CPS) are able to self-diagnose and alert technical service.

On the other hand, the virtual component consists of a digital twin, defined as the digital copy of the physical device, simulating its operation during the design and utilization phases. Digital twins represent powerful tools enabling organizations to accomplish multiple tasks. The digital twin consists of a virtual representation, serving as the real-time digital counterpart of a physical object or process. Incorporating Internet of Things (IoT) sensors into a physical object allows creation of a digital representation, thus enabling connectivity, communication, and cooperation across the supply chain. A digital twin comprises five major components: a physical system, a connection apparatus for connecting the physical and the digital twins, real-time data streams, a digital model, and an interface between the virtual object and the physical device. A network of digital twins offers organizations the possibility to connect with consumers, products, as well as assets and leverage data for the purpose of enhancing business processes. Furthermore, digital twins enable communication along the supply chain and allow organizations to build collaborative business networks in order to develop innovative business models, to enhance product performance, to maintain assets, and to distribute goods in highly competitive global markets. Internet of Things-powered digital twins contribute to improving visibility, as virtual representations provide organizations with relevant real-time insights and enable businesses to anticipate issues ahead of time. Real-time data collection serves to conduct on-going analysis, to respond to demanding customer requests, to enhance product designs, and to perform proactive as well as predictive maintenance. Digital twins serve during the product design process for the purpose of simulating the behaviour of the physical device in order to prevent errors and of supporting its implementation by mirroring user requests as well as of configuring optimal operating conditions, thus limiting costs and risks. Nonetheless, digital twins also serve to compare real and expected performances, thereby simulating distinct scenarios, to develop predictive maintenance, to reconfigure the optimal operating conditions in order to respond to changes in market demand, and to support research and development (R&D) activities [14, 41, 29]. The aforementioned functionalities are geared towards reducing operational costs, therefore enabling the pursuit of operational excellence. In addition, digital twins collect information for the purpose of identifying products most suitable for consumers, thus enabling the pursuit of product leadership as well as customer intimacy strategies.

Applications of digital twins are widely utilized across multiple industries. Multiply Labs, a robotics technology company providing autonomous manufacturing services to the pharmaceutical industry, manufactures personalized supplement capsules, custom made for each individual. Multiply Labs utilizes robotics technology for the purpose of 3D printing personalized capsules containing custom quantities of medications. Multiply Labs aims to combine all the necessary medications within one customized daily capsule.

Further components characterizing phygital operating models as well as smart factories are cyber-physical production systems (CPPS) defined as a combination of individual cyber physical systems (CPS) [42], enabling proactive and agile manufacturing systems. Furthermore, cyber-physical production systems (CPPS) enable a production whose features include the following:

- Automated, resulting in the reduction of manual labour jobs, human errors, and direct labour costs, thus guaranteeing easier integration of low-skilled or older workers, as well as the creation of new, highly skilled job profiles requiring continuous and advanced training.
- Decentralized, allowing the reduction of transport, thereby further promoting sustainability, concurrently reducing time-to-market and increasing scalability.
- Dematerialized, enabling multiple services to be delivered remotely.
- Integrated, allowing the individual enterprise to align its business processes with those of further business partners, therefore enabling supply chain extension and interconnection with high-quality services in order to manufacture 'hybrid' products and to create additional value networks.
- Focused, enabling the individual firm to focus on core activities and outsource non-core activities to partner firms.
- Collaborative, allowing the single enterprise to diversify risks and to benefit from synergies in order to capitalize on business growth opportunities and to adapt to the volatility of the markets as well as to reduce the product life cycle.
- Flexible, enabling the set of partner firms to customize products within a series production, thus engaging consumers and business partners during the design process [43].

Cyber-physical production systems (CPPS) constitute production systems consisting of multiple cyber-physical systems (CPS) and further archiving systems, in addition to those already supplied by individual cyber-physical systems (CPS), capable of sharing data for self-monitoring, self-learning, self-management, and self-adaptation. In addition, cyber-physical production systems (CPPS) represent the basis for the realization of smart factories and their interconnection with numerous actors constituting the value (eco) system. Cyber-physical production systems (CPPS) are crucial for the creation of the digital thread. The digital thread constitutes the digital representation of the entire physical life cycle of products. In addition, the digital thread consists of the digital project, evolving through consumer co-design into the digital control of manufacturing processes, in order to reduce human errors and improve quality, thus preventing production defects, resulting in the digital monitoring of product utilization or elimination. The digital thread therefore enables product digitization and traceability, thereby sharing, integrating, and converting real-time data included within business information systems into information and knowledge, which serves to maximize operational excellence in terms of productivity, procurement, as well as sustainability.

Recent investments in Industry 4.0 and digital as well as technological advancements within the manufacturing sector have contributed to revolutionizing production processes. Therefore, organizations can now largely and significantly improve resource efficiency.

Digital technologies are going to affect the modalities through which organizations innovate. Amongst future technological advancements, 5G is worthy of mention, in particular 5G technology is going to enable unprecedented levels of connectivity, upgrading 4G networks, thus providing high reliability and availability, as well as enriching the experience for specific Business-to-Consumer (B2C) applications. Combining artificial intelligence (AI), machine-learning techniques, and 5G allows to create the technical foundation for mixed reality, digital twins, as well as Industrial Internet of Things (IIoT) solutions, thus contributing to delivering the promised outcomes of advanced technologies. 5G is destined to impact multiple human activities; in addition, 5G technology is going to accelerate innovation and connect communities. Wired connectivity is going to be substituted by 5G technology, supplementing the manufacturing environment with a greater degree of flexibility. 5G appears as the major technological enabler for the digital transformation of the manufacturing industry. The present reliable connectivity technology comprises technical features designed on the basis of the direct requirements of the manufacturing business, for the purpose of automating and applying digital functions to digital processes.

3.5.1 Phygital Operating Model: Cyber Production

The phygital operating model is developed from the digital operating model previously described. The following illustration depicts the introduction of cyber plants as well as cyber production processes leading to the realization of cyber goods and services. Furthermore, the transition illustrated within Fig. 3.25 requires that organizations hire employees capable of optimizing the innovative technologies implemented in manufacturing processes. As manufacturers are increasingly transitioning towards lean as well as high-technology environments, it is critical to hire highly engaged workers capable of solving problems.

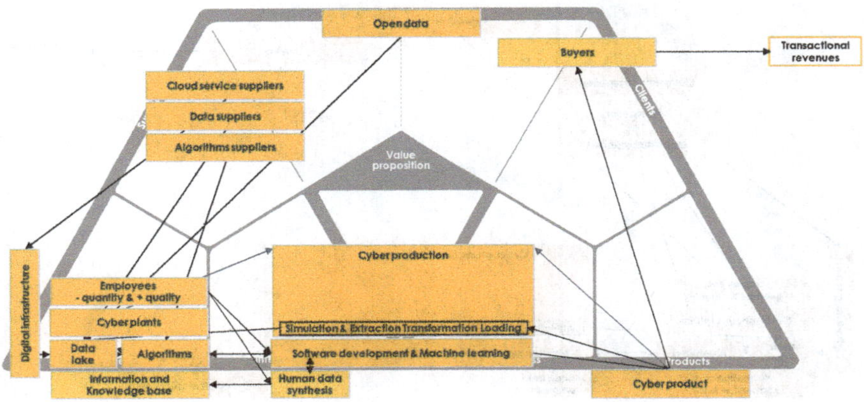

Fig. 3.25 The phygital operating model: cyber production

3.5.2 Phygital Operating Model: Cyber-Physical Production

Due to the recent advancements of information and communications technology (often abbreviated to ICT) as well as the application of industrial wireless sensor networks (often shortened as IWSN), additional technologies, including cyber-physical systems (CPS), are emerging in order to achieve effective automation. As a result, manufacturing plants, production processes, and products become cyber-physical, thus highlighting the characteristics and implications formerly identified. Cyber-physical production systems represent the integration of computation as well as physical processes. In particular, cyber-physical production systems are capable of handling immense quantities of information integrated through adequate processes. In addition, an educated, skilled, and motivated workforce represents a key prerequisite for operating a network of cyber-physical manufacturing plants (Fig. 3.26).

3.5.3 Phygital Operating Model: Configuration and Optimization

Further features characterizing cyber-physical production processes comprise configuration and optimization, as production efficiency increases, while production costs drastically decrease as a result of the introduction of advanced technologies. Optimizing computing infrastructure represents a fundamental aspect leading to improved business efficiency, as in the absence of optimized technology, organizations would not be as productive as anticipated. Digital technologies enable organizations to maintain consistent performance, thus maximizing employee productivity and minimizing the expected downtime. Technology also enhances organizational productivity, therefore simplifying time-consuming tasks (Fig. 3.27).

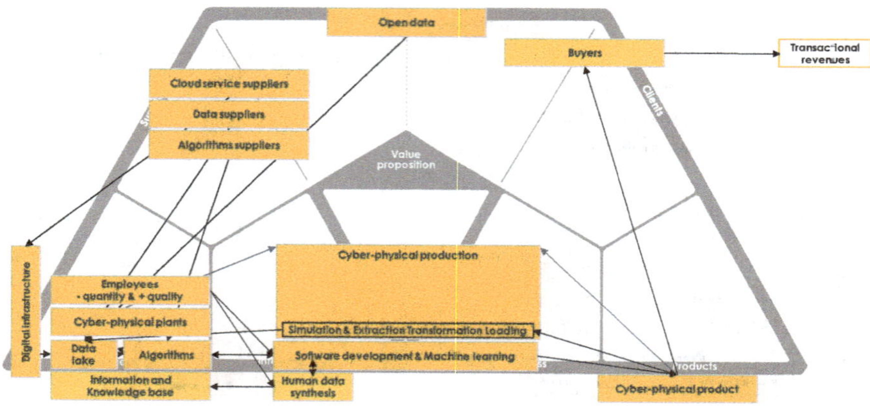

Fig. 3.26 The phygital operating model: cyber-physical production

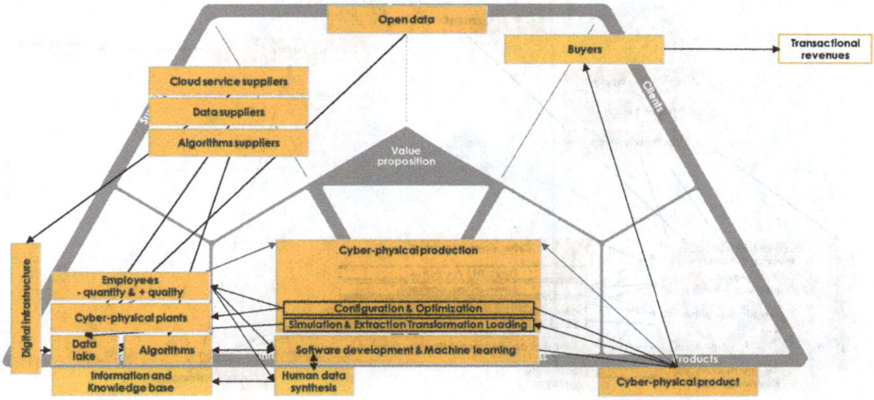

Fig. 3.27 The phygital operating model: configuration and optimization

3.5.4 Phygital Operating Model: Automation and Control

Automation and control constitute an additional significant feature and requirement, characterizing phygital operating models (as previously mentioned, production processes are automated). Furthermore, the implementation of technological advancements enables organizations to minimize system integrations issues and complexities, as well as managerial and organizational complexity, resulting in increased business efficiency (Fig. 3.28).

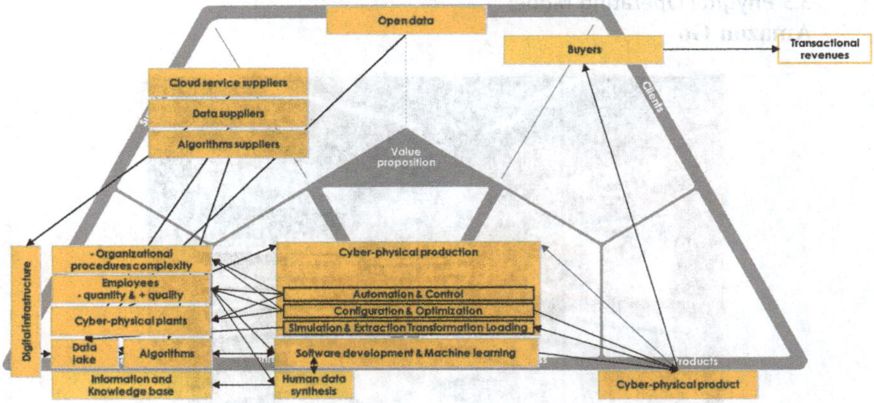

Fig. 3.28 The phygital operating model: automation and control

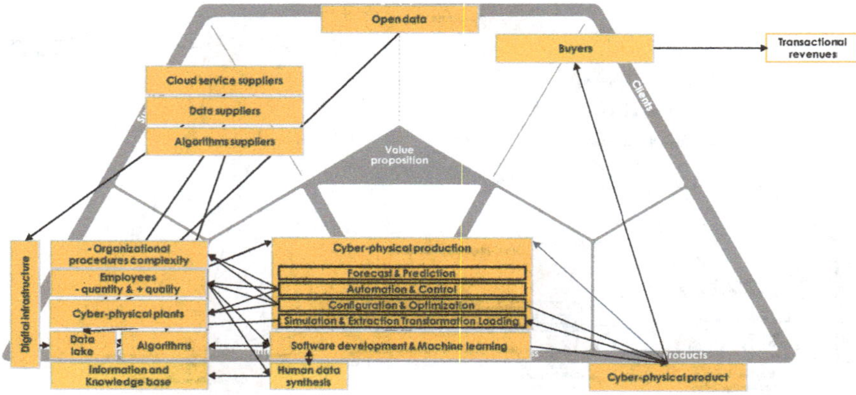

Fig. 3.29 The phygital operating model: forecast and prediction

3.5.5 Phygital Operating Model: Forecast and Prediction

Forecasting represents an effective planning tool for managing manufacturing businesses. The forecast and prediction function allows to forecast the amount of inventory and supplies necessary in order to meet changing consumer demands as well as to easily predict and detect anomalies in manufacturing processes. Furthermore, the implementation of the aforementioned digital technologies is going to enable organizations to collect real-time information across the entire supply chain and analyse data for the purpose of improving operation as well as goods through instant feedback, therefore enhancing manufacturing productivity (Fig. 3.29).

▶ **3.5 Phygital Operating Model**
Amazon Go

The phygital operating model marries the online and offline environments combining the principal aspects from each space for the purpose of enhancing consumer engagement and experience. A notable example of the online and offline convergence is represented by the Amazon Go grocery store, comprising over 20 locations within the United States. Amazon.com, Inc. presents multiple divisions, including the Amazon Go grocery store, utilizing cameras, sensors, artificial intelligence (AI), and further technologies in order to enable customers to select determined items, leave the store, and receive the receipt directly on their smartphones. Customers are required to download the Amazon Go application on their smartphone, scan a code utilizing their smartphones, select and pick up the products they wish and leave the store, without going through the traditional checkout line. Subsequently, customers will receive an electronic receipt detailing the amount of money Amazon is going to charge to their account.

3.6 Smart Factory Business Models

Smart factory business models produce an impact on the following building blocks: resources, internal processes, products, and customers. The 'intelligent' factory integrates physical as well as virtual systems and considers the entire product life cycle beginning from the product design phase to its disposal, therefore automating and dematerializing the production process in order to combine large-scale production with customization. For these reasons, smart factory business models are associated with the strategic dimension of operational excellence [42, 15, 19, 44, 45]. Operational excellence occurs once customer needs are fulfilled through the purchase of products available anytime and anywhere, thus maximizing the efficiency of production and distribution processes. It is based on a value proposition which considers the following critical success factors: price, reliability, availability, accessibility, and speed [20, 46, 18, 40, 47].

The opportunity to implement smart factory business models, as a result of the introduction of technological paradigms, is primarily characterized by the possibility of optimizing resource utilization, therefore resulting in an effective exploitation of digital resources, rendering products and processes intelligent [48–50].

Smart factories are characterized by the integration of physical as well as virtual systems, the automation and the dematerialization of production processes in order to combine large-scale production with customization, thus promoting overall performance improvement. In addition, the key components of smart factories include the following: cyber-physical systems (CPS) as well as cyber-physical production systems (CPPS), enabling the development of innovative business models for the purpose of manufacturing cyber-physical products (Fig. 3.30).

The processes characterizing smart factories can be summarized as follows:

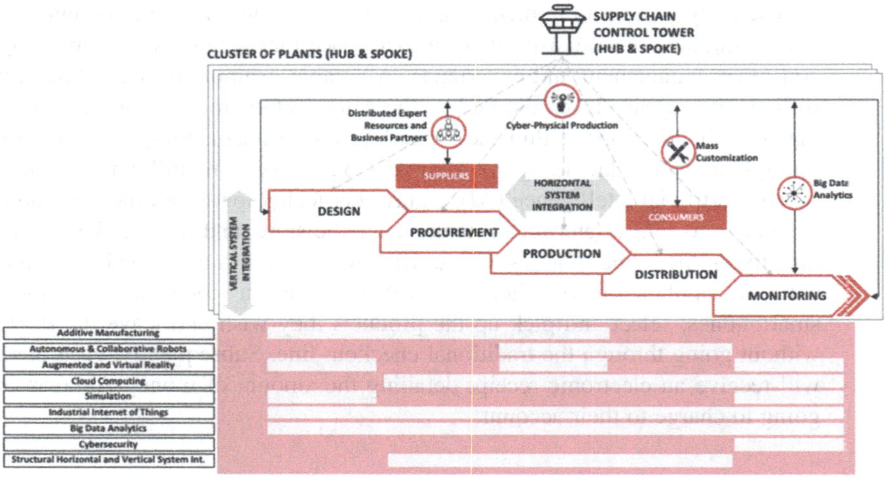

Fig. 3.30 The smart factory business model

- Computer systems are installed and configured in order to enable data analysis and the elaboration of an optimal answer so as to trigger the procurement or production of goods and services.
- Advanced systems are implemented for the purpose of rapidly generating digital goods and services, thus reducing time-to-market and autonomously optimizing production processes.
- The supply chain appears fully digitized as resources are managed efficiently and monitored in a flexible manner.
- Data are collected throughout the entire product life cycle, therefore creating relevant customer insights and adapting quickly to changing markets and events, in order to better satisfy consumer evolving needs.

Smart factories display the following characteristics:

- Connectedness. Intelligent factories represent the shift from traditional automation to an entirely connected and flexible system. Furthermore, smart factories require manufacturing processes to be connected for the purpose of generating real-time data essential to formulating decisions. Automated manufacturing enables collaboration across departments, driving greater overall supply network efficiency.
- Optimization. Smart factories present highly automated production processes, allowing operations to be executed with minimal manual intervention as well as human interaction. Automation and the synchronization of assets and equipment contribute to significantly increasing quality as well as reducing costs.
- Transparency. Data captured from processes as well as cyber-physical products are transparent, and real-time data visualizations can convert the data collected

into actionable insights. Transparent networks are crucial as transparency enables organizations to view real-time notifications and alerts, therefore formulating accurate decisions.

- Proactiveness. Smart factories enable organizational staff and manufacturing systems to anticipate potential issues or challenges rather than reacting once complexities have occurred. The capability of intelligent factories to predict future outcomes, based on historical and real-time data, contributes to enhancing uptime as well as quality and to preventing safety issues. Manufacturers can leverage multiple digital as well as physical technologies, including digital twins, enabling virtual representations to digitize manufacturing operations, thus moving beyond automation and integration and evolving into predictive capabilities.
- Agility. Advanced intelligent factories present automated and flexible manufacturing processes allowing smart factories to easily adapt to schedule as well as to product changes with minimal intervention.

The features illustrated above contribute to helping organizations decrease time-to-market as well as gaining market share and increasing profitability.

3.6.1 Smart Factory Business Model

The smart factory business model presents similar characteristics to the phygital operating model. Smart factories consist of highly digitized shop floors, continuously collecting and sharing data through automated and connected manufacturing equipment, as well as production systems. Data constitute a valuable resource characterizing the digital industrial era. Data are utilized across the organization for the purpose of self-optimizing digital devices, proactively addressing issues, enhancing manufacturing processes, and responding to shifts in demand.

The convergence of the physical manufacturing world with the digital manufacturing world enables intelligent factories to design as well as to monitor entire production processes. Smart factories implement integrated and collaborative manufacturing systems in order to render operations flexible and optimizable. Leveraging the fusion of the physical and virtual world into cyber-physical systems requires a multitude of technologies, including cloud computing as well as advanced analytics. Nonetheless, manufacturers must develop a profound understanding of digital technologies in order to unveil opportunities and to overcome challenges. Organizations willing to implement smart factory business models also achieve benefits for customers and for companies themselves.

Furthermore, lead time reduction represents a critical aspect of process enhancement, as reducing lead time contributes to enhancing delivery reliability and accuracy and enables organizations to define proper improvement actions.

In addition, digital technologies enable organizations to enhance end-product quality and reduce operational costs, thus maximizing asset utilization.

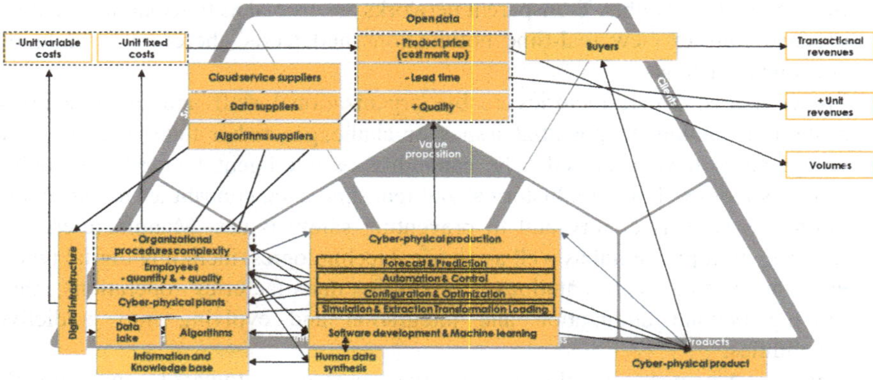

Fig. 3.31 The smart factory business model

The digital infrastructure represents a further critical component characterizing Industry 4.0, as infrastructures must deliver the computing power essential to accommodating data loads in order to conduct production monitoring [37].

Moreover, manufacturing activities are geared towards the realization of smart products consisting of cyber-physical systems providing innovative features and functions based on connectivity. Cyber-physical products display four major characteristics:

- Awareness. Smart products include sensor technology, providing access to information regarding the digital good and the surrounding environment.
- Intelligence. Digital goods are equipped with computing power, thus enabling autonomous decision-making as well as self-learning processes based on defined algorithms.
- Connectedness. Cyber-physical products enable machine-to-machine communication (often shortened as M2M), therefore facilitating interaction and automated exchange of data amongst additional cyber-physical systems, constituting the manufacturing environment.
- Responsiveness. Smart products involve control technology, resulting in autonomous product adaption based on internal and external commands [37].

Figure 3.31 pictures numerous financial implications in terms of revenues and costs.

3.6.1 Smart Factory Business Model: Boston Dynamics

Industrial robotics allows organizations operating within the manufacturing industry to increase industrial productivity, efficiency, as well as safety. In 2019, Boston Dynamics acquired Kinema Systems, an American 3D vision start-up, producing 3D vision software, enabling robotic arms to collect and deposit pallets in warehouses. Subsequently, Boston Dynamics launched the Boston Dynamics Pick system. Pick consists of a deep learning-based solution designed to operate in challenging warehouse environments. The autonomous robot integrates high-resolution 2D and 3D sensing as well as machine-learning algorithms in order to identify and locate industrial storage containers, thus eliminating bottlenecks in inbound operations.

3.6.2 Smart Factory Business Model: Decentralized Production and Control

The introduction of cyber-physical production systems (CPPS) as well as connected and distributed cyber-physical plants represent the primary components constituting the decentralized production and control business model, therefore basing the manufacturing process on advanced technologies combining both digital as well as physical elements. Ultimately, smart products are developed and obtained as a result of the process aforementioned.

Furthermore, the introduction of vertical system integration enables access to automated manufacturing equipment, therefore enabling companies to implement real-time production planning, thus drastically reducing lead time and increasing manufacturing process efficiency. The vertical 'machine-to-machine' integration involves the creation of networks, enabling constant exchange of data and information, therefore allowing business process optimization, preventing unexpected equipment failures and downtime as well as designing adaptive production control systems, through the utilization of machine-learning techniques. Leaders in digitized manufacturing comprehend the value of data generated by cyber-physical

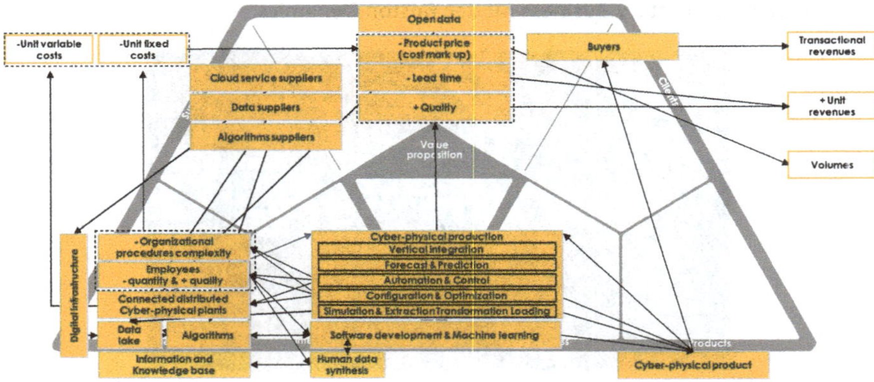

Fig. 3.32 The smart factory business model: decentralized production and control

production systems (CPPS) and exploit information accordingly. In addition, advanced analytics enable organizations to discover insights regarding performance enhancement across the manufacturing value chain, thus detecting inefficiencies based on historical data and formulating reliable hypotheses for optimization in the future. Physically capable production processes provide the possibility of manufacturing single products individually without incurring additional costs, also delivery dates are going to become increasingly reliable and operational flexibility as well as productivity are expected to rise as automated production lines are allocated efficiently. Furthermore, shop floor innovation contributes to transforming organizational production planning and control, as cyber-physical production systems (CPPS) enable decisions locally, therefore decentralizing production control (Fig. 3.32).

3.6.2 Smart Factory Business Model: Decentralized Production and Control

Pixartprinting

Digital printing describes the process aimed at printing digital-based images directly onto a variety of media substrates. Digital files, such as PDFs or desktop publishing files, can be sent directly to the digital printing press to print on paper, photo paper, canvas, fabric, synthetics, cardstock, and further substrates. The process enables both on-demand printing and lead time reduction. Pixartprinting is an Italian company, offering online printing services and focusing mainly on web-to-print services. Pixartprinting has contributed to creating a simple and straightforward ordering process, to attain the premium printing services allowing all consumers to uniquely customize their own products. During the ordering process, customers will be able to choose their preferred paper type, weight, size, shape, and additional printing features. Ultimately, the printed files are delivered to customers usually within 48 hours.

3.6.3 Smart Factory Business Model: Smart Supply Chain

Industry 4.0 has contributed to creating a disruption requiring companies to reconsider supply chain management. Connected supply chains are formed through the vertical supply network, thus allowing the seamless integration and automation of physical processes and ensuring a maximum level of transparency across the entire digital supply chain. The Fourth Industrial Revolution and the convergence of the digital and physical environments (including information technology [IT] as well as operations technology [OT]) have contributed to considerably transforming the supply chain. Smart supply networks are increasingly integrated and automated, enabled by the implementation of digital technologies as well as cyber-physical systems (CPS) [37]. Connected supply chains are amongst the essential elements for achieving operational excellence and are defined as transparent, lean, reliable, as well as cost efficient. Furthermore, smart supply chains are defined as agile, as digital supply networks contribute to facilitating the integration of changes in customer needs, therefore enabling manufacturers to become responsive to consumers [37]. Shifting from linear supply chain operations to an interconnected and open system of supply operations, frequently referred to as the digital supply network, could establish the foundation for future organizational competitiveness. In order to realize the digital supply network, manufacturers must uncover several capabilities, including structural horizontal system integration, through the multitude of operational systems powering the organization as well as vertical system integration, through connected manufacturing systems. The introduction of structural horizontal system integration coupled with vertical system integration enables the exchange of information across decentralized production sites, fostering the creation of an integrated and collaborative environment (Fig. 3.33).

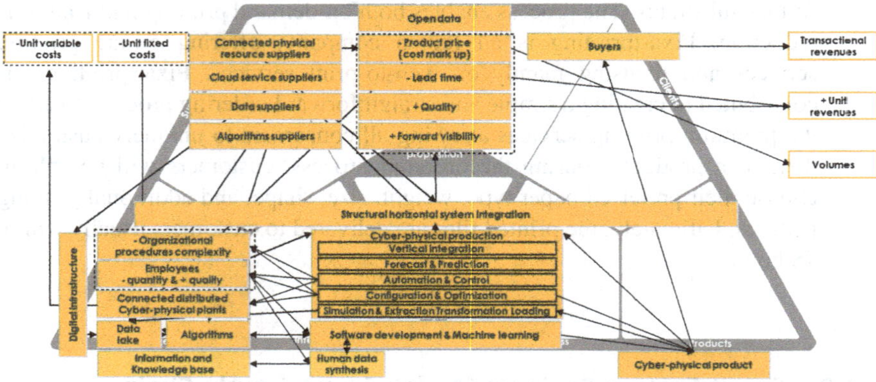

Fig. 3.33 The smart factory business model: smart supply chain

3.6.3 Smart Factory Business Model: Smart Supply Chain

Amazon.com, Inc.

Amazon represents the leading organization in the digitization of the supply chain. Amazon.com represents the largest global online retailer, selling a multitude of goods. Amazon Fresh is a subsidiary of the American e-commerce company Amazon.com, Inc. headquartered in Seattle, Washington. Amazon Fresh consists of a grocery delivery service currently available across the United States as well as in further international locations. Amazon Fresh offers grocery items for sale as well as a wide assortment of items from the main Amazon.com storefront. Items ordered on Amazon Fresh are available for home delivery within the same day or the following day, depending on the time of the order and the availability of delivery slots.

3.6.4 Smart Factory Business Model: Agile Collaboration Network

Smart factories are capable of integrating data from multiple assets in order to drive manufacturing, maintenance, digitization of operations through the application of digital twins and further activities across the entire manufacturing network, thus resulting in an efficient as well as an agile system and avoiding operational downtime. Furthermore, agility often refers to the organizational ability 'to thrive in a competitive environment of continuous and unanticipated change' [37]. Smart factories consist of flexible systems capable of self-optimizing performance across a wide network, self-adapting to, and self-learning from conditions in real-time and autonomously conducting entire manufacturing processes. Manufacturing operating models must overcome the current rigidity and respond to the business agility characterizing the recent industrial era. Agile collaboration networks describe the shift in horizontal integration towards a flexibly defined extended enterprise, allowing manufacturers to focus on core competences and enabling organizations to offer customized products in multiple markets. Local market developments as well as sudden changes in consumer demand are going to occur in Industry 4.0 and agile manufacturers are going to seize competitive advantages within the environment. Furthermore, rigid collaboration structures are going to be substituted by project-based business partnerships, essential to delivering solutions tailored to consumer needs. Business partnerships are enabled by digital technologies, fuelling collaboration while substantially decreasing complexities, through the utilization of globally decentralized resources characterizing the balanced network. In addition, access to distributed expert resources and business partners enables organizations to maximize the focus on core competencies while concurrently expanding the solution portfolio. In this regard, cloud computing solutions promote interorganizational co-operation and encourage collaboration with customers, thus creating agility in responding to environmental changes and generating business value [37].

Moreover, mass customization enables the production of small batches due to the ability to quickly reconfigure production processes in order to adapt to customer demand. Mass customization is going to enable small batch production as a result of the capability to adapt to changing consumer behaviour and expectations as well as to increase collaborative product development. Therefore, the smart factory business model combines the major elements related to the digital experience in order to deliver a comprehensive consumer experience, thus fulfilling the requirements of demanding consumers. Mass customization addresses market trends through the production of highly customized products, exploiting economies of scale, therefore maintaining a high volume of production at a reasonable low cost. In addition, mass-customized production helps to dramatically reduce lead times [15] (Fig. 3.34).

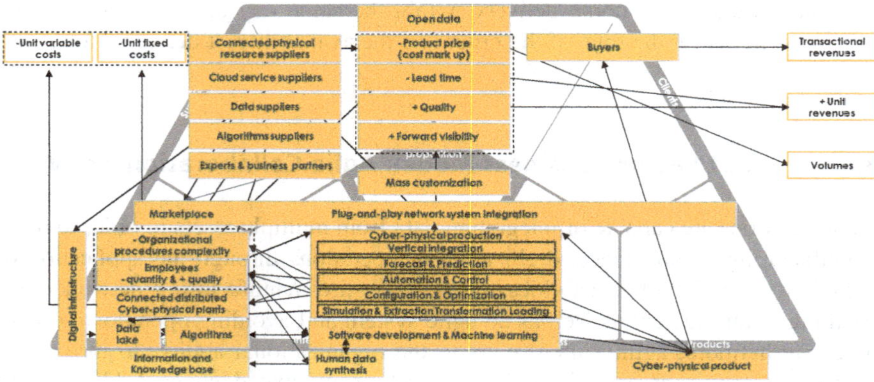

Fig. 3.34 The smart factory business model: agile collaboration network

3.6.4 Smart Factory Business Model: Agile Collaboration Network Aircraft Engine Manufacturers

Manufacturers of engines in the aerospace industry could face challenges with their external business partners, as the product life cycle management (often shortened as PLM) tools in utilization could not be sufficient in collaborating in the global extended enterprise. As a result, cloud-based collaboration platforms are implemented, thus enabling the sharing of configurations and digital product models, whilst data segmentation ensures security constraints are respected. In addition, product life cycle data are synchronized in short time intervals. Through the digitally enabled agile collaboration network, organizations are capable of shortening the time-to-market for an innovative engine generation as well as of radically reducing development costs [37].

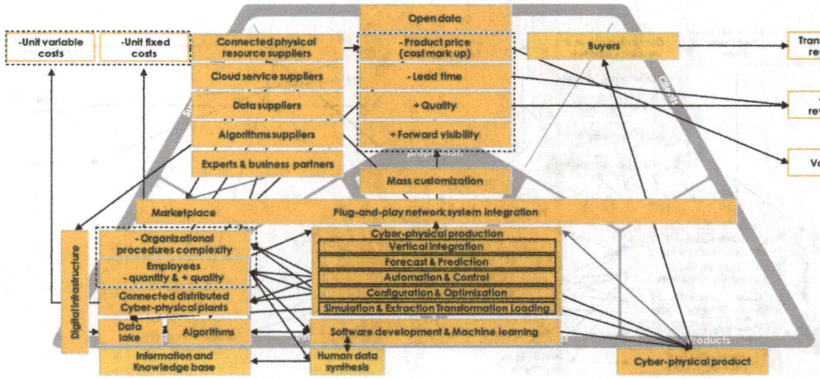

Fig. 3.35 The smart factory business mode: hub and spoke

3.6.5 Smart Factory Business Model: Hub and Spoke

Hub and spoke networks address niche markets as well as small segments of customers in remote places, thus manufacturing a small range of products in small volumes and constitute a centralized integrated logistics system designed to reduce costs. Its distribution centres receive, consolidate, and distribute products directly to various destinations. Furthermore, hub-and-spoke business models accomplish centralization, as the central hub issues commands, therefore ensuring that policies and procedures of organizations are implemented in a complete and accurate manner. From the hubs, products are distributed outwards to the spokes for further processing and distribution, thus guaranteeing efficiency [42, 15, 46, 51–55]. Figure 3.35 depicts the aforementioned process within the business model canvas.

► **3.6.5 Smart Factory Business Model: Hub-and-Spoke Exchange of Aircrafts Components**

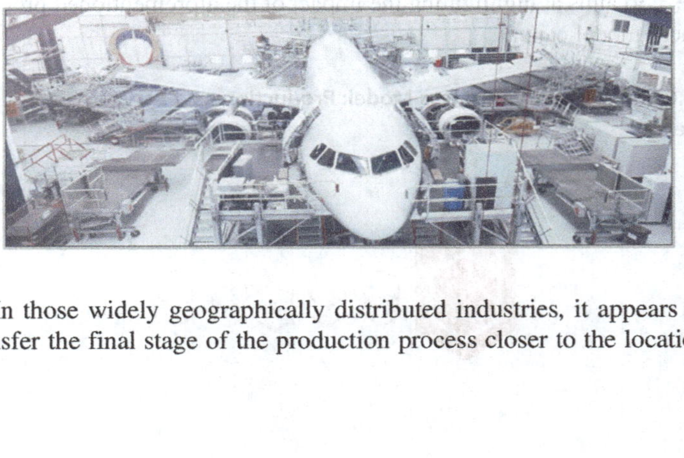

In those widely geographically distributed industries, it appears useful to transfer the final stage of the production process closer to the location where

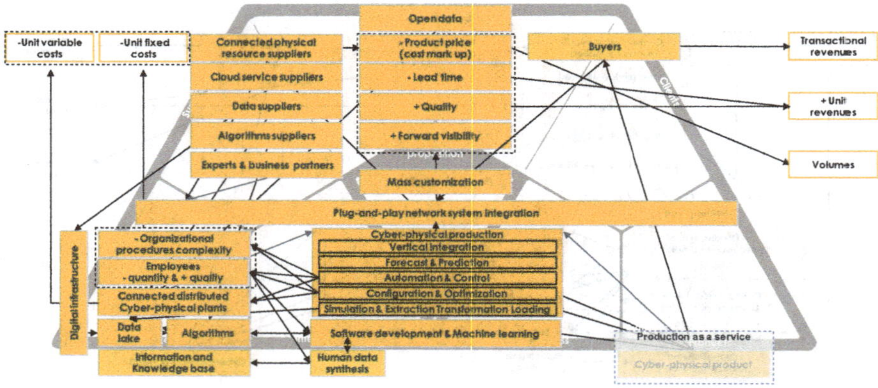

Fig. 3.36 The smart factory business model: production as a service

the object is to be purchased. The exchange of aircraft components represents the most exemplary case. Manufacturing plants are responsible for delivering aircraft components to the individual final assembly lines.

3.6.6 Smart Factory Business Model: Production as a Service

The Production as a Service business model constitutes a cloud-based, centralized framework based on a service-oriented architecture abstracting the manufacturing steps of a product as individual service requests, while connecting users with specific customized small batch manufacturing needs. Furthermore, organizations utilize a constant stream of data deriving from connected operations as well as production systems in order to learn and adapt to evolving demands and to address shifting priorities.

Figure 3.36 aims at highlighting the impact of the aforementioned process within the business model canvas.

▶ **3.6.6 Smart Factory Business Model: Production as a Service**
Beelse

Beelse is a French start-up founded in 2016. The French start-up provides cloud-based manufacturing industry solutions. Beelse aims at supporting organizations in the integration of digital manufacturing technologies into their production processes. Production as a Service enables enterprises to automate and to dematerialize inventories as well as to manufacture on-demand in multiple locations in order to enable companies to locate manufacturing facilities closer to the emerging customer base. The solution employs cloud manufacturing in order to offer customers control and insights into the production process. Furthermore, on-demand manufacturing contributes to reducing lead time as well as overhead costs.

3.7 Servitization Business Models[1]

Servitization business models produce an impact on the following building blocks: products, resources, internal processes, and customers. Servitization involves the exploitation of the numerous opportunities offered as a result of the introduction of digital technologies in order to integrate, complement, or replace the production of an asset with that of services.

Servitization business models are associated with the strategic dimension of product leadership. Product leadership occurs once customer needs are fulfilled through the purchase of customized products that respond to the specific requests of a niche of users. It is based on a value proposition which considers the following critical success factors: material and immaterial qualities and performances [23].

In the recent years, researchers have distinguished between organizations manufacturing products and enterprises providing services. Nonetheless, a well-defined distinction between the process models proposed does not appear to exist. The widespread diffusion of digital technologies, including the introduction of artificial intelligence (AI) as well as robotics are rendering the traditional distinctions questionable, thus altering the competitive dynamics amongst firms [56].

Research on servitization has proliferated in recent years alongside the interest of the scientific community towards the topic. Servitization is defined as the transformational process through which organizations transition from a product-centric to a service-centric business model and logic. The term 'servitization' was coined by Professor Sara Vandermerwe and Professor Juan Rada in 1988 in reference to a movement attributable to an increasing offering of 'fuller market packages of "bundles" of customer-focused combinations of goods, services support, self-service and knowledge' [57]. The seminal contribution is considered the birth of the

[1] The present sub-chapter is based on Bagnoli, C., Massaro, M. and Messina, M. (2022). Chapter 15: Organizational change towards servitization: guiding manufacturing firms through the digital transformation process. In M. J. Sousa, P. S. Kumar, F. Dal Mas &S. Sousa (Eds.), *Advancements in Artificial Intelligence in the Service Sector*. Routledge, London.

Table 3.1 Distinctions between product and service

Product	Service
Object	Activity or process
Tangible	Intangible
Separation of production and consumption	Simultaneity of production and consumption
Homogeneous	Heterogeneous
Can be stored (non-perishable)	Cannot be stored (perishable)
Less difficult to determine quality	More difficult to determine quality
Ownership	Temporary possession

servitization literature which has been continually enriched with additional contributions from further scholars.

Five streams of research have been identified in order to highlight distinct aspects relating to servitization.

1. First, marketing research has stressed the difference between the technical characteristics distinguishing products and services. Services differ from goods along four dimensions: intangibility, heterogeneity, inseparability, and perishability, often abbreviated to IHIP. In addition, services involve relationships between service providers and consumers. Furthermore, goods are tradable, and users can reap economic benefits from the ownership.

Table 3.1 illustrates the principal distinctions characterizing products and services.

2. Second, service management scholars underline the impossibility of applying manufacturing operations concepts to service industries, as service companies possess particular characteristics. Furthermore, service innovation requires tailored innovation and custom-designed services as well as the adoption of innovative guiding principles and a definite value structure (comprising the servitized value chain, the organizational structure, operational resources, and digital capabilities) and alignment between the servitization strategy and the organizational design.
3. Third, Professor Andy Neely affirms that servitization implies the innovation of organizational capabilities and processes for the purpose of generating value shifting from selling goods to selling product-service systems [58]. The current stream of research has underlined the role of value chains as well as the optimization of service networks. In this regard, traditional manufacturing firms have transitioned from product manufacturers to providing desired outcomes to customers. Business strategies based on product-service system establish a value proposition focused on end user needs, therefore enabling the design of a need-fulfilment system with lower impacts, in terms of environmental and social benefits [59, 1].

4. Fourth, although product-service systems represent the next frontier for innovation, scholars note that it appears complex to translate consumer needs into technical features.
5. Finally, services are defined as a system of interacting components including individuals, technologies, and businesses in order to drive innovation as well as competition through value co-creation. As a result, innovative solutions are emerging as advanced instruments in order to enable collaborative consumption of goods and services, following trends of sustainability and dematerialization [56].

Multiple recent literature reviews on servitization illustrate numerous concepts utilized in order to describe the phenomenon. Researchers often mention the following expressions when referring to servitization and related concepts: product-service systems (often shortened as PSS), transition from products to services, dematerialization, and integrated solutions [60].

Initially, servitization related to companies operating in all industries [57]. Nevertheless, currently servitization has become synonymous of service growth within manufacturing organizations [61]. Indeed, the diffusion of advanced technologies leading to the digitalization of production processes has contributed to accelerating the combination of manufacturing as well as service functions within organizations, thus enabling industrial enterprises to design additional services or suggesting businesses to enrich existing goods in order to differentiate the value offering. Nonetheless, complexities in performing the transition include the required cultural change for a product-centred organization to convert into a service-oriented, therefore integrating and building internal and external competences, as well as reconfiguring organizational resources and activities in order to address evolving environments.

The classical distinction between goods and services has been gradually substituted by the emerging service-dominant logic, defining services as the process of generating something for external parties and goods as the instruments providing digital services. Furthermore, the service-dominant logic focuses on the importance of operant resources, generating skills as well as knowledge rather than operand resources, valuing production, and transactions. The involvement of customers, constituting the value network, represents an additional aspect characterizing the service logic [56].

Although servitization is studied with respect to the industrial sector, the provision of services encompasses all economic sectors. The service expansion presents a double effect. On the one side, services constitute an additional source of revenues. On the other side, services serve as a means to gain consumer loyalty, thus contributing to guaranteeing customers a unique experience as well as to building long-term customer relationships, while establishing recurring revenue [56].

The literature cites a variety of organizations implementing servitization business models. Nonetheless, amongst the companies operating in the service industry, Rolls-Royce represents the most notable example, as the corporation delivers power-by-the-hour as opposed to transferring ownership of the gas turbine engines to airline companies, thus offering engine as well as accessory replacement service on a fixed-cost-per-flying-hour basis. The solution aligns the interests of manufacturers and operators, as customers only pay for engines performing well. Furthermore, the service enables operators to remove risk related to unscheduled maintenance events [56, 62].

Manufacturing as well as service companies are required to implement the following actions (constituting the major drivers of servitization business models) for the purpose of achieving a successful servitization process: building durable customer relationships (potentially leading to a higher degree of customer retention as well as trust), reducing environmental costs, extending existing service offerings, and optimizing asset utilization.

Servitization represents a competitive tool in order to create dependency, to differentiate the market offering, to diffuse innovation, and to enhance research and development (R&D) activities. Furthermore, due to technological advancements, organizations can offer an extensive portfolio of services at the lowest cost, while reaching consumers worldwide. Servitization also represents a valid option in order to allow manufacturers to avoid imitation, therefore disclosing innovative sources of competitive advantage. Literature identifies various benefits deriving from servitization, influencing production as well as consumption patterns. The strategic perspective suggests that adding services to the portfolio enables firms to retain customers, as servitization places consumers at the centre of the value offering, thus benefitting from highly customized offerings and contributing to increasing consumer satisfaction. Moreover, services can be utilized in order to prolong the life cycle of products or goods can be shared and utilized jointly.

Nevertheless, literature distinguishes amongst several challenges manufactures encounter during the transition from goods to services. The principal challenge is represented by the servitization paradox. Firms encounter difficulties in the generation of incremental profits through services. Despite the increase in revenue, servitized companies encounter lower net profits due to the increased costs deriving from the implementation of the service strategy. Servitization failure originates as most organizations do not possess the adequate capabilities and therefore recruit specialized talent in order to compete, or due to difficulties 'connected to the knowledge transfer, adaptation and integration between the solution provider and the customer' [58]. In order to overcome the service paradox, organizations should recur to accounting practices, utilizing performance measurement systems, and distinguishing between service adoption (identifying the proportion of customers adopting the service) and service coverage (establishing whether the proposed solution fulfils consumer requests). An additional alternative to overcoming the challenges hindering innovative service development requires organizations to evaluate the alignment amongst the three dimensions of value offering, value structure, and value network. In addition, scholars argue that not all services are equal, since digital services can either substitute products (cannibalization) or facilitate the sale of goods. Furthermore, the present challenge requires a related change in the corporate mind-set as well as the implementation of innovative business models and the introduction of advanced technologies, thus contributing to the emergence of digital services based on data generated through user–product interactions. Nonetheless, servitized companies encounter further challenges linked to the cost of the provision of digital services: once enabled, the marginal costs of digital services amount to zero, consequently customers perceive services as less valuable [63, 1].

Digital technologies are influencing competition amongst organizations, thus contributing to forging relationships between upstream and downstream players. Embracing technology is crucial to enabling digital servitization. Digitalization requires organizations to adopt a transformation model comprising four stages. The initial phase views companies as manufacturers providing product-related services. The second stage requires organizations to produce information and technology (often shortened as IT) services. Subsequently, manufacturers distribute pure digital services. The fourth and final stage entails that firms create digitalized product-service systems (PSS), thus providing additional service bundles [56].

Servitization has been influenced by the advancement in information and communication technologies (often abbreviated to ICT) contributing to maximizing the value of goods as well as to integrating product and service development. Digital servitization involves multiple technology-enabled business models, enabling organizations to seize competitive advantages through the provision of customer knowledge-based digital services, throughout the entire product life cycle. Internet of things (IoT) technologies enable companies to realize innovative and previously unconceivable services as well as to monitor usage metrics.

Professor Arnold Tukker identifies three established categories of product-service systems business models, including product-oriented, use-oriented, and result-oriented [64]. An additional classification concerns distinct paths servitized businesses might follow in the transition from basic to advanced services. The transition path implies that firms abandon basic services in favour of advanced solutions. On the other hand, the extension path suggests organizations to maintain both basic as well as advanced services. Servitized companies differ in terms of their value offering, as services can be classified according to distinct forms:

- Basic services include standardized services, requiring a low customer involvement.
- Intermediate services require a slight customer involvement.
- Advanced services comprise highly customized solutions, demanding a strong interaction between suppliers and customers [56].

In addition, customer-related capabilities represent an additional broad set of skills to be acquired by companies. Moreover, scholars recommend enterprises to involve customers in co-creation processes and to adopt open innovation in order to accelerate service innovation. Customers are defined as co-creators of value, as consumers interact throughout the entire project life cycle. Nonetheless, customer involvement can be limited to the initial and the final stages of the project as well. As a result, organizations must communicate with customers as well as integrate consumers in their processes.

Collaboration represents the key in order to pursue innovation and increase performance. In this regard, the open innovation paradigm has been extended to servitized firms as well. Companies must consider converting to open-service businesses and evaluate the role of external stakeholders in the development of innovative service scenarios. Nevertheless, firms generally face a further challenge referred to as the commodity trap, deriving from goods easily available worldwide as well as low customer switching costs, therefore encouraging consumers to change suppliers. Consequently, servitized organizations aim at achieving a sustainable differentiation, thus designing high-quality services at the least cost [56].

As Professor Elena Casprini states, the value chain of a servitized enterprise presents a cyclic structure. Furthermore, the value chain considers the core activities adding value, including:

- **Problem identification**. Prospective customers present unsolved issues. Thus, companies compete in order to uncover unidentified consumer needs. The capability of revealing user needs offers organizations the opportunity to obtain market shares as well as to seize competitive advantage.
- **Value offering development**. Once enterprises succeed in identifying consumer needs, businesses must provide solutions in order to solve the issues. In the servitization context, the answer resides within the value offering. Digitalization distinguishes between hardware and software components of the service offering. The components enable multiple functions ranging from the collection of considerable quantities of data to the provision of additional services.
- **Value offering monitoring**. The value offering performance monitoring represents the final step, enabling enterprises to increase customer satisfaction through continuous feedback, to enrich the value offering through the design of innovative services and to discover consumer needs in order to propose additional solutions [56].

Research has contributed to formulating a classification relating to servitization business models. Servitization business models are classified into (Fig. 3.37):

- Product ownership, referring to the traditional operating model consisting of a single transaction between two parties, transferring the ownership of a physical product in exchange for a lump-sum payment.
- Product + service add-on hardware, representing the evolution of the traditional operating model through the sale of innovative connected services, thus integrating digital technologies. Furthermore, add-on hardware services aim at enhancing product performance.
- Product + service add-on software, including service solutions enabled through product utilization.
- Product availability, transforming the classical one-time transaction model into a recurring source of revenue. Therefore, not limiting consumer relationships to a single transaction.

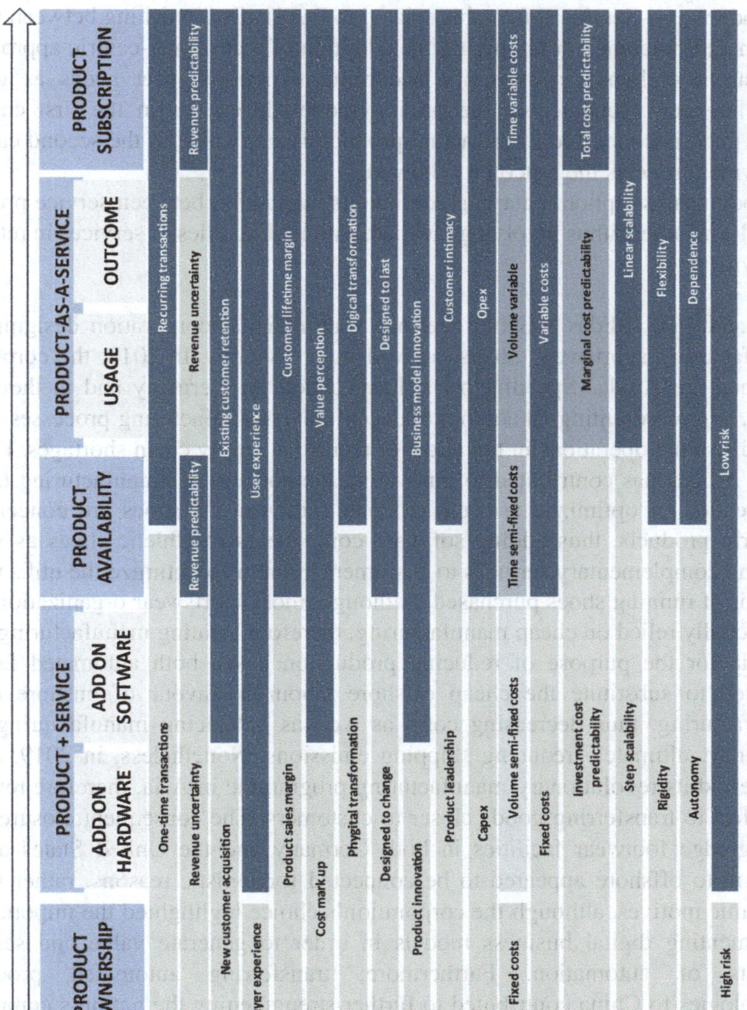

Fig. 3.37 The servitization business model

- Product-as-a-Service (often abbreviated to PaaS), distinguishing between utilization and outcome, and implying the adoption of a customer-centric approach, as customers choose to subscribe according to specific user needs as well as preferences. Nonetheless, the pricing model differs, as in the first case, the payment is connected to the actual product usage, whilst in the second case, the payment reflects the outcome achieved.
- Product subscription, entailing continuous interaction between service providers and consumers, thus involving the distribution of bundles of services in return for payment.

In addition, Adidas AG is a German multinational corporation designing and manufacturing sportswear, shoes, as well as accessories. In 2018, the corporation launched the Adidas Speedfactories, located both in Germany and in the United States, and representing an attempt to decentralize manufacturing processes as well as to distribute operations in the midst of recurring supply chain shortages. Furthermore, Adidas has contributed to innovating the concept of manufacturing through the creation of optimized as well as smart footwear. Its shoes are conceived as platform products, thus adding software components to athletic shoes as well as offering complementary services to customers in order to optimize the utilization of the pair of running shoes purchased. Although, most sportswear organizations have traditionally relied on cheap manufacturing, therefore locating manufacturing plants in Asia for the purpose of reducing production costs, both automated factories intended to substitute the cheap offshore labour in favour of onshore robotic manufacturing, thus decreasing costs as well as enhancing manufacturing cycle efficiency, ultimately reducing shipping emissions. Nonetheless, in 2019, Adidas transferred its revolutionary manufacturing programme in Asia, therefore reversing the effort to transferring goods closer to customers. The consequent closure of the cutting-edge footwear facilities in both Germany and the United States and the decision to offshore appeared to be connected for logistic reasons, rather than to economic motives, although the corporation's choice highlighted the importance of implementing digital business models in order to generate value and seize the benefits of automation. Furthermore, transferring automated production technologies to China contributed to further strengthening the nation's competitive advantage [65].

3.7.1 Servitization Business Model: Product Ownership

The servitization business model originates from the phygital operating model and the smart factory business model. Product ownership business models represent traditional operating models involving a one-time transaction between service providers and customers. Manufacturers transfer product ownership to consumers in return for payment.

Figure 3.38 depicts the aforementioned process within the business model canvas.

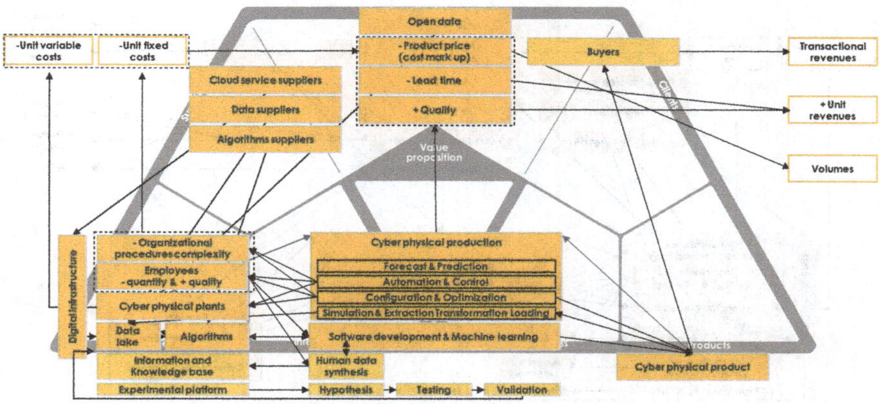

Fig. 3.38 The servitization business model: product ownership

3.7.1 Servitization Business Model: Product Ownership –Ford Motor Company

Ford Motor Company represents an American multinational auto manufacturer and constitutes an example of product ownership business model. Ford is currently implementing the traditional value chain model, thus internalizing multiple activities preceding the sale of its vehicles. The relationship with the customer is articulated around a one-time transaction, involving the ownership of the good, in exchange for payment.

3.7.2 Servitization Business Model: Product + Service (Add-on Hardware)

The servitization business model is further classified into product + service add-on hardware and add-on software. The add-on hardware business model contributes to evolving the traditional operating model, offering technology-enabled service solutions. Digital services promise enhanced product performance, thus increasing revenue streams. Solutions include dematerialized goods as well as additional services. Servitization involves offering additional services in order to supplement goods for the purpose of growing the business. Organizations implement

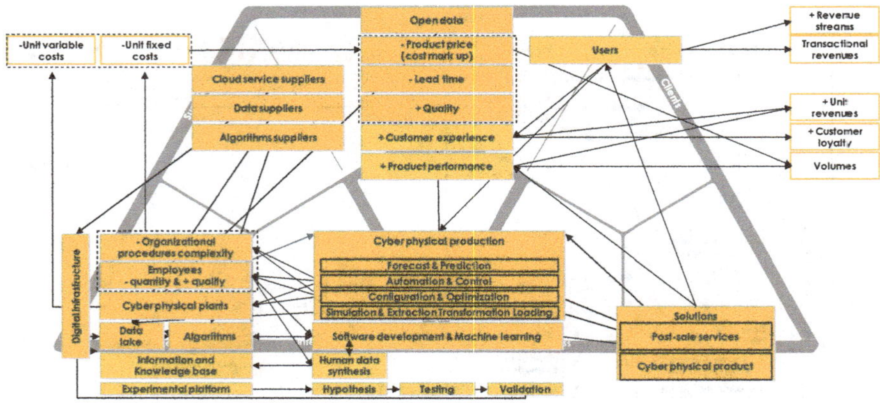

Fig. 3.39 The servitization business model: product + service (add-on hardware)

servitization business models as services guarantee sustainable and predictable cash flow generation as well as increased customer satisfaction. Furthermore, organizations increase efforts in order to forge direct relationships with users and collect data and insights as a result of interactions. In addition, service providers maintain ownership of cyber-physical products.

Figure 3.39 illustrates the aforementioned process within the business model canvas.

3.7.2 Servitization Business Model: Product + Service (Add-on Hardware)

Schneider Electric SE

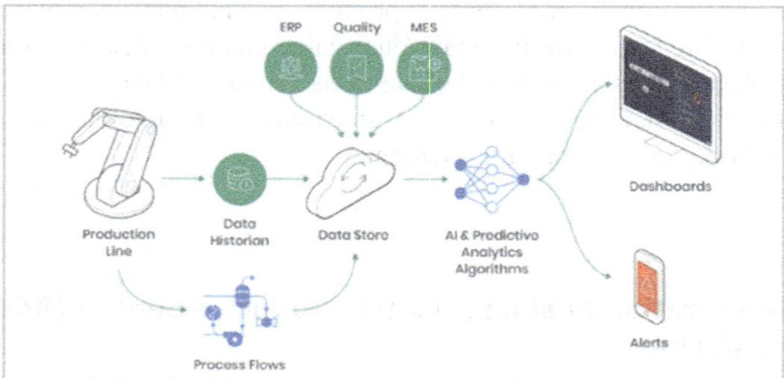

Schneider Electric SE represents a multinational organization designing digital solutions for the purpose of achieving efficiency as well as sustainability. Schneider Electric's predictive maintenance technologies enable facility managers to understand the current state of their electrical distribution equipment. Monitoring the condition of equipment provides trending data in order to help anticipate and plan maintenance activities.

Potential issues can be identified and corrected ahead of time rather than resulting in costly, unplanned downtime. Schneider Electric's predictive maintenance solutions help reduce downtime as well as improve workplace safety.

3.7.3 Servitization Business Model: Product + Service (Add-on Software)

The add-on software business model indicates a service solution designed to be utilized in conjunction with the related dematerialized good. Digitalized product-service systems (PSS) are defined as integrated bundles of tangible goods, intangible service solutions, and digital architectures designed to fulfil individual customer requirements through automated operation, thereby enhancing consumer outcomes. Cyber-physical products and related services are customized, therefore enabling organizations to develop direct relationships with consumers, thus avoiding intermediaries and revealing increased revenue growth as well as additional business opportunities. In order to servitize, it appears fundamental to cooperate with stakeholders, including suppliers as well as final customers. In addition, customers are crucial in assessing the quality of service solutions. Manufacturers experiencing de-commoditization obtain strategic differentiation via business model innovation. Business model innovation enables organizations to address market challenges as well as needs. Nonetheless, delivering service solutions often requires organizations to re-consider the internal organizational structures and processes as well as the external networks. Qualified employees (including IT and finance professionals) are also critical to the successful implementation of servitization [56, 1]. Furthermore, technological innovation enhances service design, translating into increased performance and reduced cost.

Figure 3.40 depicts the aforementioned process within the business model canvas.

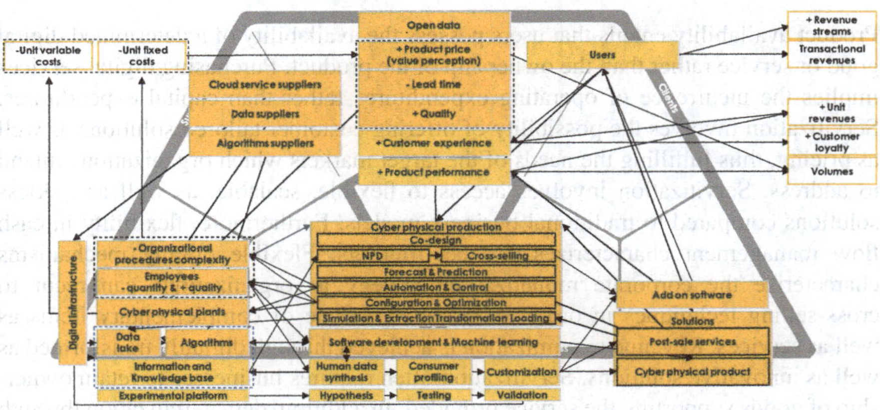

Fig. 3.40 The servitization business model: product + service (add-on software)

► **3.7.3 Servitization Business Model: Product + Service (Add-on Software)**

Apple Inc.

The App Store consists of a digital distribution platform designed and maintained by Apple Inc. for mobile applications on its iOS and iPadOS operating systems, enabling organizations as well as individual developers to build and distribute their own applications. The store allows users to browse and download applications developed within its digital infrastructure. Applications aim at providing services, enabled through the utilization of digital devices, including smartphones.

3.7.4 Servitization Business Model: Product Availability

Product availability entails that users possess the availability of a determined digital good or service rather than the ownership of the product. Purchasing digital services implies the incurrence of operating expenditures rather than capital expenditures. Servitization involves the possibility of offering customer tailored solutions as well as pricing, thus fulfilling the needs of the target markets which organizations intend to address. Servitization involves access to flexible, scalable, as well as riskless solutions compared to traditional business models. Furthermore, flexibility in cash flow management characterizes service offerings. Flexible pricing mechanisms characterize the corporate monetization strategy as organizations can recur to cross-selling techniques in order to increase the sale of complementary items as well as services. Revenue maximization is achieved through digitally transformed as well as innovative solutions. Servitization often requires businesses to retain ownership of goods supporting the service provided. In addition, dematerialization through

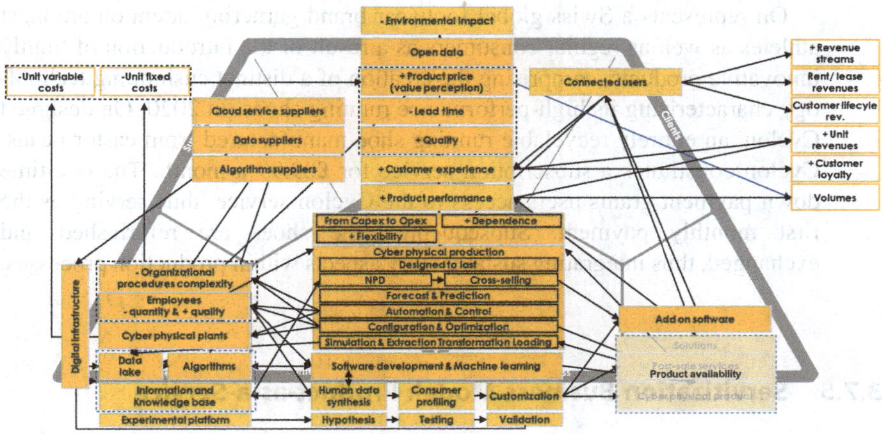

Fig. 3.41 The servitization business model: product availability

services represents a proposal for decreasing environmental impacts. Therefore, firms are incentivized to realize products designed to last longer, thus reducing material usage as well as the corporate environmental impact.

Figure 3.41 depicts the aforementioned process within the business model canvas.

▶ 3.7.4 Servitization Business Model: Product Availability

On

On represents a Swiss global footwear brand garnering attention amongst athletes as well as regular consumers as a result of the introduction of highly innovative products, comprising the creation of a distinct cushioning technology characterizing its high-performance running shoes. In 2020, On designed Cyclon, an entirely recyclable running shoe manufactured from castor beans. Cyclon constitutes a subscription service for £29.99 a month. The one-time down payment grants users access to the Cyclon service, thus serving as the first monthly payment. Subsequently, the shoes are refurbished and exchanged, thus integrating sustainability aspects within production processes.

3.7.5 Servitization Business Model: Product-as-a-Service

Product-as-a-Service (often shortened as PaaS) business models imply that service providers own the product throughout its entire life cycle and transfer the good to customers as a service, thus contributing to changing the related pricing model. Although the aforementioned service solutions constitute disruptive as well as differentiating options, Product-as-a-Service (PaaS) business models require organizations to develop adequate skills as well as digital capabilities. The current solution is intended from its inception to be sold as a service. Customers pay in exchange for the utilization of the product, thus generating on-demand revenues as a result of the on-going relationships with consumers, whilst providers maintain ownership of the dematerialized good. Furthermore, linear scalability (included within the value proposition) refers to the ability to increase production inputs by a certain percentage in order to obtain a corresponding percentage increase in outcome. In addition, pay-per-use and pay-per-outcome constitute the principal pricing models adopted by servitized companies, implementing a Product-as-a-Service (PaaS) business model, and therefore contributing to reducing total costs.

Figure 3.42 illustrates the aforementioned process within the business model canvas.

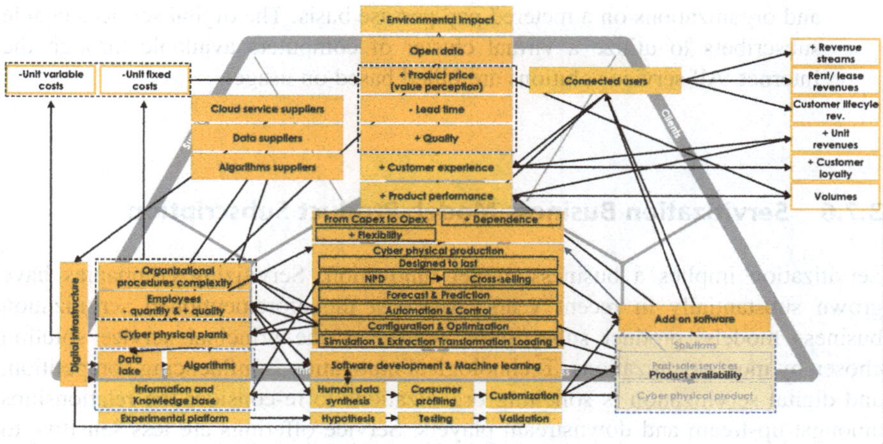

Fig. 3.42 The servitization business model: product-as-a-service

3.7.5 Servitization Business Model: Product-as-a-Service

Amazon.com, Inc.

Product-as-a-Service (PaaS) business models require users to pay in order to obtain access to the determined virtual product or to achieve an outcome. Amazon Web Services represents a subsidiary of Amazon.com, Inc. responsible for the provision of on-demand cloud computing platforms as well as application programming interfaces (often shortened as APIs) to individuals and organizations on a metered pay-per-use basis. The digital services enable subscribers to utilize a virtual cluster of computers available through the Internet. All service solutions are billed based on usage.

3.7.6 Servitization Business Model: Product Subscription

Servitization implies a business model innovation. Servitizing companies have grown substantially in recent years. Amongst the aforementioned servitization business models, product subscription represents the principal service solution chosen by most organizations. Technological innovation is influencing competition, and digital servitization is soliciting organizations to re-consider the relationships amongst upstream and downstream players. Service offerings are less sensitive to price competition and tend to provide significant margins worldwide compared to traditional business models. Companies investing in subscription services are rewarded with higher revenue growth, as recurring revenue represents the foundation of subscription-based organizations [56, 1]. Manufacturers must design an innovative system in order to enable the innovative distribution of advanced services. Technologies enable the development of further metrics, which organizations adopt in order to establish a proper pricing model as service models differ from the traditional sale of goods. Servitization implies a shift from one-time transactions based on selling branded products to distributing outcomes and experiences, thus forging deep relationships with customers. Furthermore, service organizations are characterized by a shift from cost-plus pricing (describing a pricing model according to which the selling price is determined through the introduction of a certain mark-up percentage to an individual product's unit cost) to customer lifetime value profitability through a dynamic, value-based pricing (reflecting a pricing strategy involving the application of prices according to the estimated value of a good or service to the customer). The aforementioned pricing method contributes to enhancing profitability, thus generating greater prices. In addition, organizations no longer plan their activities based on historical data, instead due to the introduction of subscription models and the collection of data, businesses utilize forward-looking statements in order to plan the future distribution of service solutions. Therefore, businesses tend to focus on delivering efficient, consumption-based services.

Figure 3.43 depicts the aforementioned process within the business model canvas.

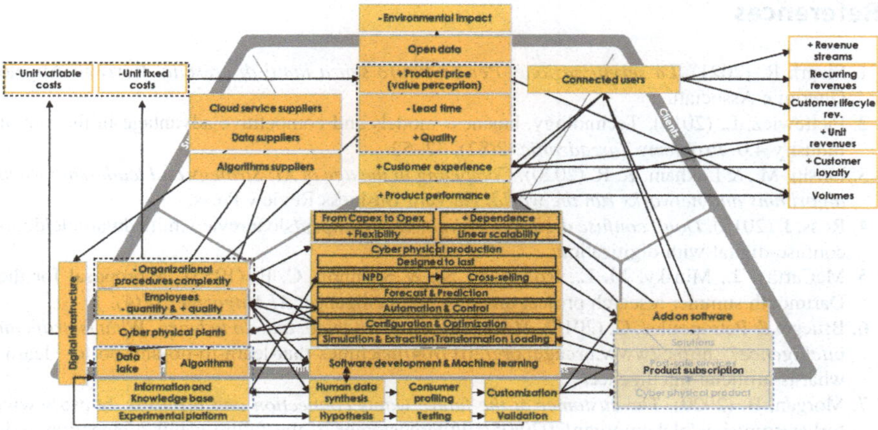

Fig. 3.43 The servitization business model: product subscription

3.7.6 Servitization Business Model: Product Subscription

The Walt Disney Company

The Walt Disney Company constitutes an American multinational mass media and entertainment conglomerate, operating five major business segments, including Disney Media and Entertainment Distribution (often abbreviated to DMED). The division's responsibilities include the management of the corporation's direct-to-consumer businesses, comprising its streaming services. Disney+ represents an on-demand streaming service operated by the Media and Entertainment Distribution division of The Walt Disney Company. On Disney+, subscribers can watch a variety of Disney movies and television series utilizing their digital devices. Furthermore, Disney+ requires the payment of a fixed monthly subscription fee.

References

1. Siagri, R. (2021). *La servitizzazione. Per un futuro senza limiti di crescita*. Edizioni Angelo Guerini e Associati.
2. Pietrewicz, L. (2019). Technology, business models and competitive advantage in the age of Industry 4.0. *Problemy Zarzadzania, 2*(82), 32–52.
3. Iansiti, M., & Lakhani, K. R. (2020). *Competing in the age of AI: Strategy and leadership when algorithms and networks run the world*. Harvard Business Review Press.
4. Ross, J. (2017). *Don't confuse digital with digitization*. https://sloanreview.mit.edu/article/dont-confuse-digital-with-digitization/
5. McCarthy, J., Minsky, M. L., Rochester, N., & Shannon, C. E. (1955). A proposal for the Dartmouth summer research project on artificial intelligence. *AI Magazine, 27*(4).
6. Bruegel & Petropoulos, G. (2017). *Machines that learn to do, and do to learn: What is artificial intelligence?* https://www.bruegel.org/2017/04/machines-that-learn-to-do-and-do-to- learn-what-is-artificial-intelligence/
7. Morgan, B. (2020). *The customer of the future wants connection and progress*. https://www.forbes.com/sites/blakemorgan/2020/05/19/the-customer-of-the-future-wants-connection-and-progress/?sh=7458cad41171
8. Baban, A., Cirrincione, A., & Mattiello, A. (2017). *Mind the change: capire il cambiamento per progettare il business del futuro*. Guerini Next srl.
9. McClean, T. (2021). *Machines are learning from each other but it's a good thing*. https://www.forbes.com/sites/forbestechcouncil/2021/02/03/machines-are-learning-from-each-other-but-its-a-good-thing/?sh=3662b6894c5
10. Morning Consult. (2020). *The fastest growing brands of 2020*. https://morningconsult.com/fastest-growing-brands-2020/
11. Mattiello, A., et al. (2019). *Cromosoma e innovazione: 20 visioni dal MIT per capire come la tecnologia riscrive la genetica delle organizzazioni*. Guerini Next srl.
12. Nonaka, I., & Takeuchi, H. (1995). Knowledge-Creating Company. *Knowledge-Creating Company*, no. December 1991, 3–19.
13. Montanus, M., & Thesis, G. (2016). *Business models for Industry 4.0*. Delft University of Technology.
14. Cotteleer, M. J., Mahto, M., & Sniderman, B. (2016). *Industry 4.0 and manufacturing ecosystems: Exploring the world of connected enterprises*. Deloitte University Press.
15. Breunig, M., Kelly, R., Mathis, R., & Wee, D. (2015). *Industry 4.0-how to navigate digitization of the manufacturing sector*. McKinsey & Company.
16. Geissbauer, R., Schrauf, S., & Vedso, J. (2016). Industry 4.0: Building the digital enterprise. *Web*.
17. Bremicker, M., & Heynitz, H. v. (1990). *The factory of the future*.
18. C. F. I. (2015). *CFI, Roadmap per la ricerca e l'innovazione - Research and Innovation Roadmap*.
19. C. Consulting. (2016). Implications of Industry 4.0 for CIOs.
20. Treacy, M., & Wiersema, F. (1993). Customer intimacy and other value disciplines. *Harvard Business Review, 71*(1), 84.
21. Bechtold, J., Bernhofer, L., Kern, A., & Lauenstein, C. (2011). Industry 4.0: The Capgemini consulting view. *Capgemini Consulting Digital Transformation of Supply Chains*.
22. Baur, C. (2015). *Manufacturing's next act*. http://www.mckinsey.com/business-functions/operations/our-insights/manufacturings-next-act
23. Geissbauer, R., Koch, V., Kuge, S., & Schrauf, S. (2014). Industry 4.0: Opportunities and challenges of the Industrial Internet. PricewaterhouseCoopers. *Aktiengesellschaft Wirtschaftsprüfungsgesellschaft, 2*(3), 52.
24. Akeson, L. (2016). *Industry 4.0: Cyber-physical systems and their impact on business models*.
25. Labory, S., & Bianchi, P. (2018). *The fourth industrial revolution*. National Academies Press.
26. Wixom, B. (2014). *Data monetization survey*.

27. Gawankar, S. A., Gunasekaran, A., & Kamble, S. S. (2018). Sustainable Industry 4.0 framework: A systematic literature review identifying the current trends and future perspectives. *Process Safety and Environmental Protection, 117*, 408–425.
28. Chiang, R. H. L., Grover, V., Liang, T. P., & Zhang, D. (2018). Creating strategic business value from big data analytics: A research framework. *Journal Management Information System, 35*(2), 388–423.
29. Heo, J. Y., Lim, C., Kim, K. H., Kim, K. J., Kim, M. J., & Maglio, P. P. (2018). From data to value: A nine-factor framework for data-based value creation in information-intensive services. *International Journal Information Management, 39*, 121–135.
30. Corradi, A. et al. (2018). Smart appliances and RAMI 4.0: Management and servitization of ice cream machines. *IEEE Transactions on Industrial Informatics*.
31. Chen, C. H., Lin, T. J., Xu, X., & Zheng, P. (2018). A systematic design approach for service innovation of smart product-service systems. *Journal of Cleaner Production, 201*, 657–667.
32. McKinsey & Company. (2021). *Five technology building blocks to power customer-centric, data-driven growth*. https://www.mckinsey.com/industries/consumer-packaged-goods/our-insights/five-technology-building-blocks-to-power-customer-centric-data-driven-growth?cid=other-eml-alt-mip-mck&hdpid=6829c48b-75ff-4cc5-8074-ff68fdd32765&hctky=11943466&hlkid=63edb78442f94de7a7befc55621a03c4
33. Henning, W., Johannes, H., & Wolfgang, K. (2013). *Recommendations for implementing the strategic initiative Industrie 4.0*.
34. Choudary, S. P., Parker, G. G., & Van Alstyne, M. W. (2016). *Platform revolution: How networked markets are transforming the economy and how to make them work for you*. W. W. Norton & Company, Inc.
35. Choudary, S. P. (2015). *Platform scale: How an emerging business model helps startups build large empires with minimum investment*. Platform Thinking Labs Pte. Ltd.
36. Gunther McGrath, R. (2016). *Is your company ready to operate as a market?* https://sloanreview.mit.edu/article/is-your-company-ready-to-operate-as-a-market/
37. Bechtold, J., Bernhofer, L., Kern, A., & Lauenstein, C. (2014). *Industry 4.0 – The Capgemini consulting view: Sharpening the picture behind the hype*. Capgemini Consulting.
38. Mancin, M., Russo, S., & Vaia, G. (2019). L'azienda. In C. Bagnoli (Ed.), *Economia aziendale tra tradizione e innovazione* (pp. 45–57). Giuffrè Francis Lefebvre.
39. Buliga, O., Müller, J. M., & Voigt, K. I. (2018). Fortune favors the prepared: How SMEs approach business model innovations in Industry 4.0. *Technology Forecasting & Social Change, 132*, 2–17.
40. Burmeister, C., Luettgens, D., & Piller, F. T. (2015). Business model innovation for industrie 4.0: Why the 'industrial internet' mandates a new perspective on innovation.
41. Manyika, J., et al. (2015). The Internet of Things: Mapping the value beyond the hype. *McKinsey Global Institute*, no. June, 144.
42. GTAI. (2013). *Smart manufacturing for the future*.
43. Burke, R., Hartigan, M., Laaper, S., Mussomeli, A., & Sniderman, B. (2017). *The smart factory: Responsive, adaptive, connected manufacturing*. Deloitte University Press.
44. Mrugalska, B., & Wyrwicka, M. K. (2017). Towards lean production in Industry 4.0. *Procedia Engineering, 182*, 466–473.
45. Berger, R. (2016). *Joint-production 4.0. nuove sfide per la cooperazione economica italo-tedesca*.
46. PricewaterhouseCoopers. (2015). *Digital manufacturing. Cogliere l'opportunità del rinascimento digitale*.
47. Economico, C., & Giobbe, S. (2017). Industry 4.0: cosa ne pensano le aziende italiane.
48. Grosvenor, R., Liu, Y., & Qin, J. (2016). A categorical framework of manufacturing for industry 4.0 and beyond. *Elsevier, 52*, 173–178.
49. The Boston Consulting Group. (2015). *Industry 4.0: The future of productivity and growth in manufacturing industries*.

50. Heppelmann, J. E., & Porter, M. (2015). How smart, connected products are transforming companies. *Harvard Business Review*, no. October, 1–38.
51. Kleer, R., Piller, F. T., & Weller, C. (2014). Business models with additive manufacturing. *Advances in Production Technology*.
52. Carlberg, M., Kreutzer, S., Moeller, C., & Smit, J. (2016). *Industry 4.0: Study for the ITRE Committee*.
53. Koch, M., Merkofer, P. D., & Schlaepfer, R. C. D. (2014). Industry 4.0 challenges and solutions for the digital transformation and use of exponential technologies. *Deloitte*.
54. Blanchet, M., Rinn, T., Thaden, G. von, & Thieulloy, G. de (2014). *The new industrial revolution: how Europe will succeed*.
55. Cassani, P. (2017). *Gli hub & spoke produttivi modello di business hub & spoke*.
56. Casprini, E. (2019). *Beyond servitization: New managerial challenges for manufacturing firms*. Wolters Kluwer.
57. Vandermerwe, S., & Rada, J. (1988). Servitization of the business: Adding value by adding services. *European Management Journal, 6*(4).
58. Valtakoski, A. (2017). Explaining servitization failure and deservitization: A knowledge-base perspective. *Industrial Marketing Management, 60*.
59. Annarelli, A., Battistella, C., & Nonino, F. (2019). *The road to servitization: How product service systems can disrupt companies' business models*. Springer.
60. Velamuri, V., Neyer, A. K., & Moslein, K. (2011). Hybrid value creation: A systematic review of an evolving research area. *Journal fur Betriebswirtschaft, 61*, 3–35.
61. Kowalkowski, C., Gebauer, H., & Oliva, R. (2017). Service growth in product firms: Past, present and future. *Industrial Marketing Management, 60*, 82–88.
62. Boehm, M., & Thomas, O. (2013). Looking beyond the rim of one's teacup: A multidisciplinary literature review of Product-Service Systems in Information Systems, Business Management, and Engineering & Design. *Journal of Cleaner Production, 51*, 245–260.
63. Vendrell-Herrero, F., Bustinza, O., Parry, G., & Georgantzis, N. (2017). Servitization, digitization and supply chain interdependency. *Industrial Marketing Management, 60*, 69–81.
64. Tukker, A. (2004). Eight types of product-service system: Eight ways to sustainability? Experiences from Supronet. *Business Strategy and the Environment, 13*(4), 246–260.
65. Bagnoli, C. (2019). *Industria 4.0, rivoluzione a metà senza nuovi modelli di business: l'innovazione tecnologica di prodotto da sola non basta, ci vuole quella strategica*. https://www.ilsole24ore.com/art/industria-40-rivoluzione-meta-senza-nuovi-modelli-business-ACDkCS1?refresh_ce=1

The Integration of Digital Business Models: The Amazon Case Study

4

Amazon was founded in 1994, when Jeff Bezos, its founder, realized that the advent of the Internet would have represented the event contributing to revolutionizing the world. From that moment onwards, Amazon shifted from a modern online bookshop to the number one online retailer in the world. Its vision is to be an everything-un-store, therefore innovating traditional shopping, and capable of offering an unlimited assortment of products, becoming "a place where customers can come to find and discover anything and everything they might want to buy online" [1], just like the Amazon River—the biggest "river" selling and distributing products all over the world, whose long reach contributes to overcoming all obstacles it encounters.

The path to the foundation of the business has been indeed challenging; nonetheless, the strength and the vision of its founder led to its success, against all odds and despite unfavourable forecasts. The success of Amazon is also associated with the multiple strategic innovations that enabled the business to reach the highest position ever in the most prestigious rankings of innovation performance, therefore confirming its progressive transformation within the past decade. Also, the strategy that led to such success is not based on its successful competitive positioning but rather on an approach referred to as "competitive movement". It is a movement that Amazon applies relentlessly. Over the past year, the company has evolved its strategy which led to an unexpected move: the opening of its first department store [2].

Nowadays, competition within the digital era involves two major issues: unlimited choice and customer attention. Amazon has built its competitive advantage as a result of the implementation of digital technologies related to Industry 4.0. It has integrated them developing an organic organizational structure, characterized by strategic innovation. Amazon's current value proposition involves the reconfiguration of all four business models characterizing Industry 4.0: smart factory, data-driven, servitization and platform (Fig. 4.1).

Amazon was among the first companies in the world to understand that data represent such a valuable and strategic asset in order to generate value and design

C. Bagnoli et al., *Digital Business Models for Industry 4.0*, Future of Business and Finance, https://doi.org/10.1007/978-3-030-97284-4_4

Fig. 4.1 Amazon: integrating and reconfiguring all four digital business models

new product and service offerings both in the digital world and in the real world. The vision of Bezos for Amazon is to create a business focusing on data [1].

Amazon represents indeed a successful example of algorithmic business as it is powered by data and algorithms and as it learnt from its mistakes and its experiments operating on a strategic and organizational level, as an advanced machine learning system, both when the priorities for Bezos included dimensional growth and building up a competitive critical mass (Get Big Fast strategy) and when it became important to realize a profit, for economic and financial sustainability reasons (get your house in order strategy).

Amazon is a multinational technology company that has introduced the figure of the chief algorithms officer and that has developed algorithms in various contexts, in order to automatically create editorial contents of its product sheets within its website (the Amabot algorithm represents one of the first innovations that contributed to achieving scalability), for realizing online price comparison (pricing bots), for transforming the order fulfilment process and for logistic automation, for reconfiguring the layout of its warehouse (Kiva Robotics), for tracking customers with motion detection inside Amazon Go's new physical stores and so forth. Amazon is governed through the adoption of a complex system of algebraic equations and algorithmic structures that contribute to continuously enhancing its performance.

Amazon is probably the only company, on a global scale, that exploits to its advantage the principal "tensions" characterizing our time: the industrial robots and automation replacing human labour (man vs. machine); the capitalist model of augmented firm emerging from the digital transformation which no longer results sustainable on a systemic level in the medium long term (profit vs. sustainability); the big companies smothering small businesses (artisan production vs. industrial production); and the companies possessing data having the power of controlling and manipulating the system (knowledge exploration vs. knowledge exploitation).

How was it possible for Amazon to develop over time all of this, in such a complex and articulated manner? How did it adopt a digital business model? Indeed, Bezos can take credit for the success of his own business initiative. Nonetheless, Amazon has always resorted to open innovation, albeit with logic and results that have evolved over time, also by virtue of the experience accumulated due to a long list of failures and successes. Bezos, who is passionate about books and reading, was always careful to grasp useful insights for developing his own business and is familiar with *The Innovator's Dilemma* and Christensen's theory on disruptive innovation. From the beginning, he was perfectly aware that although Amazon represented a real disruption for traditional chain stores, at the same time, it had to protect itself against possible disruptions coming from the external organizational environment. For these reasons, Amazon initially resorted to open innovation investing millions of dollars in start-ups and dot-com companies with the aim of identifying and experimenting with potential new marketplaces and product categories on the Internet. However, when the dot-com bubble burst in 2000, these investments appeared disastrous because of the huge losses recorded. A few years later, Amazon changed its approach without ignoring the application of Christensen's advice from a practical point of view. In order to avoid, for example, that e-books replace paper books, Bezos created an autonomous organization in Palo Alto, California, far away from Seattle to design and develop what would later become the Kindle e-book reader, instructing his team to "kill bookstores", as Amazon itself did at the time. The same modus operandi would be utilized in the development of further smart products and the AWS infrastructure. Furthermore, he acquired multiple companies (Zappos, Kiva Robotics, Diapers.com, Whole Foods, etc.) for the purpose of achieving his strategic goals. A book entitled *Good to Great* by Jim Collins made Bezos understand the importance of moving from a strategic point of view: first of all, deciding and focusing on what Amazon is good at and then managing the virtuous circles in an effective manner and acquiring new companies for the purpose of accelerating their own competitive advantage logics and strengthening synergies (the so-called Amazon flywheel), rather than creating new ones [1].

Amazon also collaborates with various universities and research centres in the United States and in the rest of the world. Seattle was chosen as its headquarters because of this while also benefiting from its proximity to Microsoft and to an ecosystem having the potential to develop innovative projects in the digital field. However, few people know that due to his knowledge and appreciation of the theory of open innovation, Bezos became one of the first supporters and pioneers of crowdsourcing, personally investing in the following initiatives consisting of a Cambrian Ventures Internet incubator and despite having received a denial from the board of directors to invest as Amazon. This led to the introduction of another pioneering initiative referred to as the Mechanical Turk project that in turn laid the foundation for the Amazon Web Services (AWS).

Following the evolution of Amazon, we can identify the following four major steps, which are traceable to the classification of digital business models illustrated in the previous chapters:

1. Amazon as an e-commerce and distributor (smart factory business model)
2. Amazon as a marketplace platform (platform business model)
3. Amazon as a service provider (data-driven business model and servitization business model)
4. Amazon as a phygital product/store platform (data-driven business model and platform business model)

The following paragraphs describe each of these evolutionary steps with the intention of highlighting, in a non-exhaustive manner, how Amazon has managed to grasp the opportunities of the Fourth Industrial Revolution (Fig. 4.2).

4.1 Amazon as an E-Commerce Business and Distributor (Smart Factory Business Model)

Amazon today is intensely focused on developing itself as a cyber-physical system, an intelligent network of machines and computers, an advanced ICT system for enhancing the customer experience and intelligently managing its internal processes and supply chain (Fig. 4.3).

Amazon's value proposition is to be the "the unstore" [1] that can sell anything.

Being an unstore means that Amazon is free rather than bound, like its competitors, by the traditional (and deemed obsolete) rules or customs of doing business in the commerce sector. The only constraint, which is also Amazon's obsession, is what is deemed best for the customer. The following design choices initially characterized the unstore: limitless shelf space, personalized service for each customer, allowing both positive and negative reviews, placing used products alongside new ones so customers can make informed choices and aiming to offer low prices every day and excellent customer service (cost and customer intimacy strategies).

Selling "anything" has led to constantly broadening and expanding the product categories available on its e-commerce. The first has been that of books. Even if obviously not all of them have always met Amazon's expectations (e.g. the production and sale of personalized jewellery and costume jewellery), today the categories comprise many other products, and its target market includes B2C as well as B2B customers.

In order to develop its value proposition, Amazon has configured itself as a smart factory. Its business model is characterized by the interconnection of its own physical infrastructures and machines and its own digital systems in an intelligent network which extends along the value chain. Upstream, next to its suppliers and partners in order to pursue maximum efficiency, and downstream, including transport logistic operators for deliveries and especially customers, for a pervasive and persistent customer experience, thanks also to the introduction and wider diffusion of proprietary proactive smart "cyber" products (Kindle e-book reader, Fire tablet, Amazon TV, Echo and Alexa).

Fig. 4.2 Amazon's business model evolution

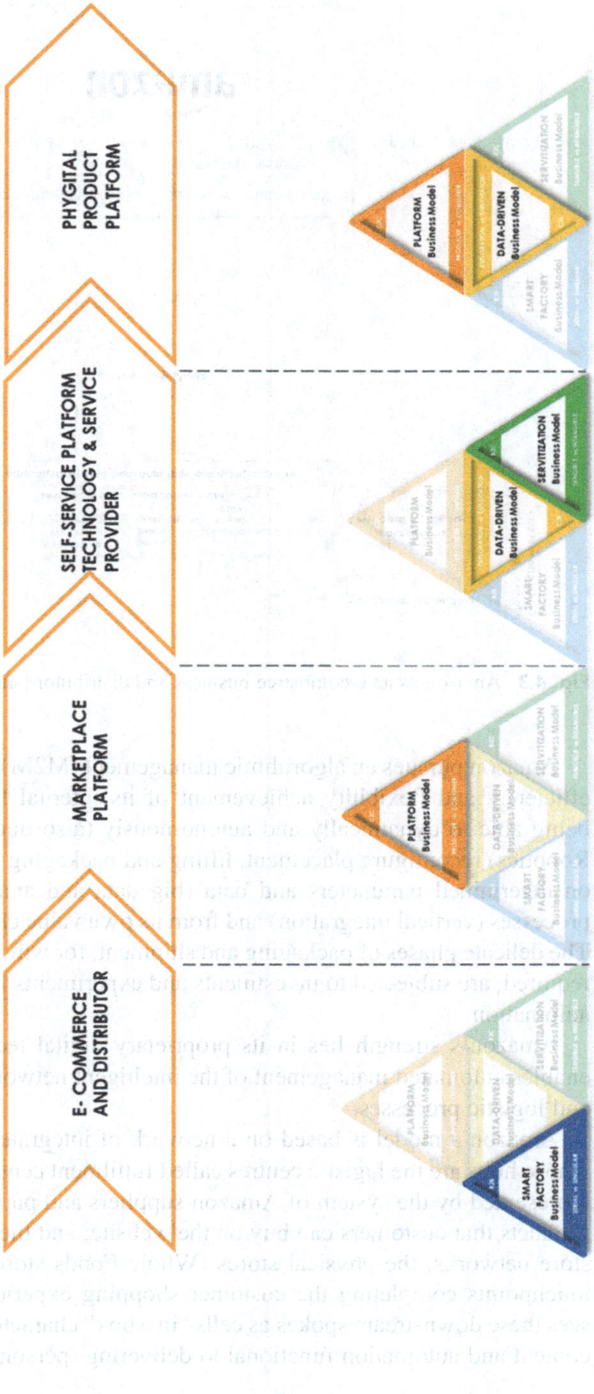

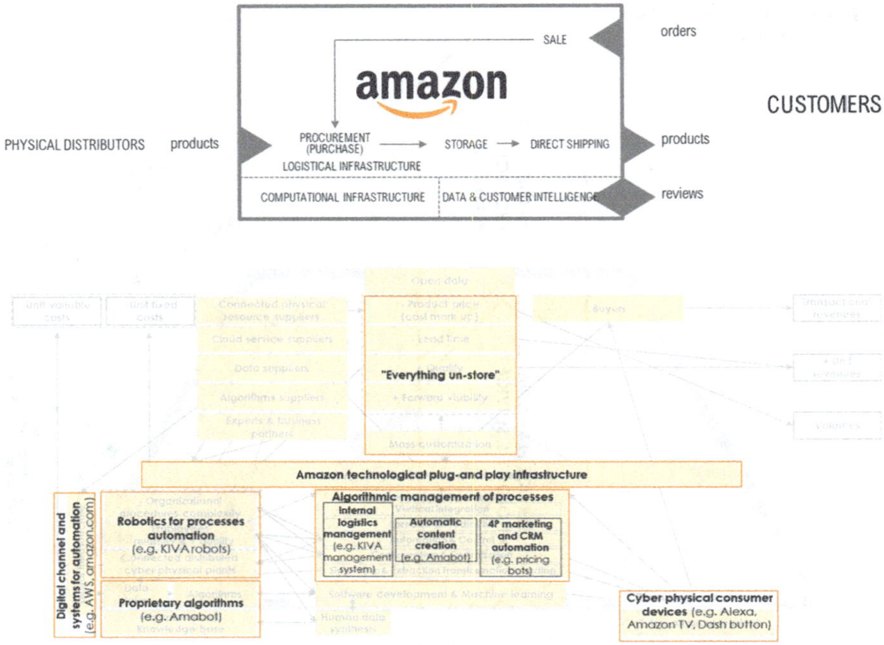

Fig. 4.3 Amazon as an e-commerce business and distributor (smart factory business model)

Amazon pursues an algorithmic management (M2M) oriented towards maximum efficiency and flexibility achievement of its internal logistic processes, therefore being able to dynamically and autonomously (also due to the acquisition of Kiva Robotics) reconfigure placement, lifting and packaging goods to be delivered based on determined parameters and data (big data and analytics) deriving from other processes (vertical integration) and from its own value chain (horizontal integration). The delicate phases of packaging and shipment, for which human intervention is still required, are subjected to investments and experiments for the purpose of increasing automation.

Amazon's strength lies in its proprietary digital technology infrastructure that enables automated management of the intelligent network connecting its production and logistic processes.

Amazon's model is based on a network of integrated logistic hubs and spokes, whose hubs are the logistic centres called fulfilment centres; the upstream spokes are represented by the system of Amazon suppliers and partners for the procurement of products that customers can buy on the website; and the downstream spokes are the store networks, the physical stores (Whole Foods stores, Amazon Go) that act as touchpoints completing the customer shopping experience. The present approach sees these downstream spokes as cells "in a box" characterized by high technological content and automation functional to delivering (personalized) Amazon services.

IoT technologies and sensors, artificial vision systems and recognition of customer profiles and contactless payments constitute solutions already adopted by Amazon and subject to constant enhancements. It seems instead that it will take some time before we see these spokes become production cells for on-demand and local distribution of goods. Several years ago, Bezos personally invested in the three-dimensional printer start-up called MakerBot, but in a note to the shareholders sometime later, he also declared that this technology was (and still is) just interesting and promising and that it would be able to fully establish itself as a means to manufacture finished consumer products in the distant future [3].

A further experimental initiative involves the development of intelligent drones (technology: autonomous robots) for goods home delivery (the so-called last mile problem), although their adoption, due to regulatory and legislative constraints, is still far off.

The adoption of the smart manufacturing business model enabled Amazon to minimize delivery times from weeks to days, to a few hours, and steadily increase the level of service offered to its customers (an unbridgeable competitive advantage for many nowadays).

Amazon has built a significant share of its success on mass customization. In terms of production, considering Amazon's business model, we refer to the provision of services, rather than the production of goods. The (proprietary) products—smart devices—that it creates are functional to conveying the services offered.

Since the very beginning, Amazon has focused on developing proprietary algorithms that allow it to automate and scale operational processes.

Among the fundamental algorithms utilized during the beginning, we find Amabot, for the personalization of content on the Amazon.com website, which allowed Amazon to replace editorialists [1].

An additional innovation involves product recommendation techniques (books first and foremost) that might attract customer attention, based on purchase history (interests), preferences and web browsing history (latent interests) as well as on customer behaviour and preferences (big data analytics).

A further brilliant and innovative solution introduced by Amazon consists of the possibility for its users/customers to write product reviews on the website, which constitutes an aspect that allows for mass customization (thus considering product marketing, promotion and communication).

The same algorithmic definition of pricing, following the use of bots that automatically scan and map the prices applied by competitors, makes it possible to improve the product offering, therefore ensuring the lowest price on the market, since it seems to be a critical and primary element influencing consumer decisions.

The production of the service is therefore conceived and realized for mass customization.

Amazon's current emphasis on the production of smart products, in particular Amazon Echo and Alexa devices, can be traced back to the desire to personalize the interaction with the user by proposing a virtual assistant for the digital shopping session that can be always present during everyday life, an assistant who continuously learns and exploits the opportunities of artificial intelligence.

The Kindle e-book reader itself, in a certain sense, is part of the mass customization of paper books, as it allows to customize the graphic format (e.g. font size) and the fruition of contents (e.g. dictionary for searching specific terms, texts available in different languages) adapting them to user needs. Amazon has therefore introduced the possibility for authors to print their works on demand, connecting the trend of self-publishing and self-production, to that of mass customization. Through the Encore programme, authors received 70% of the cover price of e-books to be uploaded on the Kindle Store. The CreateSpace service instead made it possible to print a single book on demand at the time of ordering on Amazon.com, effectively bypassing its own suppliers.

4.2 Amazon as a Marketplace Platform (Platform Business Model)

Amazon became a business platform the moment it introduced Amazon Marketplace, that is, the possibility for third-party retailers to sell their products on Amazon.com. At first glance, it may appear a controversial and counter-intuitive choice, and Amazon buyers were doubtful to allow competitors to join the platform. In reality, it turned out to be a far-sighted move that Bezos had thought about and whose implications can be examined through the digital business models. Amazon has since been able to further expand its product assortment, continuing to fulfil customer needs and also involving competitors finalizing the sale in its place. Nonetheless, in doing so, it renewed its promise to the consumer that it was the digital store that could offer the widest selection of products at the most affordable prices, therefore making the marketplace appear partially transparent following the entry of third-party sellers (Fig. 4.4).

The Amazon Marketplace is considered B2B2C as it facilitates the connection of third-party sellers with customers and end-consumers. The former includes sellers, who can sign up to the platform with basic or pro-accounts, and vendors—who can only access via invitation from Amazon itself. Sellers have at their disposal the seller central platform and can "exploit" Amazon as an online store in order to sell their products (these appear on Amazon with the following indication: "Sold by seller . . ."). Vendors, on the other hand, can also exploit Amazon as an actual distributor (products appear as "Sold and shipped by Amazon").

Amazon initially built much of the competitiveness characterizing its business model on logistics management. Following the adoption of the platform model, it made two core solutions available to sellers.

The basic one consists of the Fulfilled By Merchant (FBM) solution, which requires the seller to prepare and ship the order received and managed through Amazon, a mechanism that is similar to the one that would have applied on the seller's own e-commerce or e-Bay. Amazon then simply charges the seller a commission based on the sales price of the item.

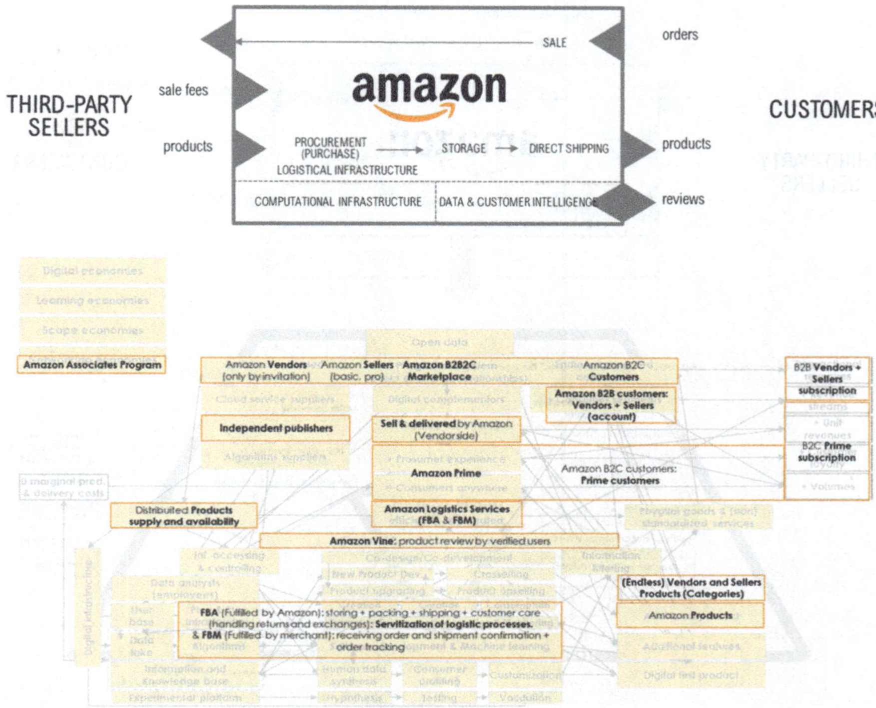

Fig. 4.4 Amazon as a marketplace platform (platform business model)

The second solution consists of Fulfilled By Amazon (FBA) and represents the integration of the servitization business model and marketplace platform, which is discussed in the subsequent paragraphs.

On the other hand, such evolutions enabled Amazon to further feed its huge data reservoir, both in quantitative and qualitative terms. This is because it is able to observe the behaviour of competitors themselves, their communication choices and their pricing and promotional policies, as well as which new products and categories to suggest. It is also able to observe and understand data in order to evaluate the sales performance and the interest of customers. At that point, it can decide to sell a determined product or a particular product category directly. Amazon has therefore integrated in a synergic and systematic way its own business model also considering the data-driven paradigm.

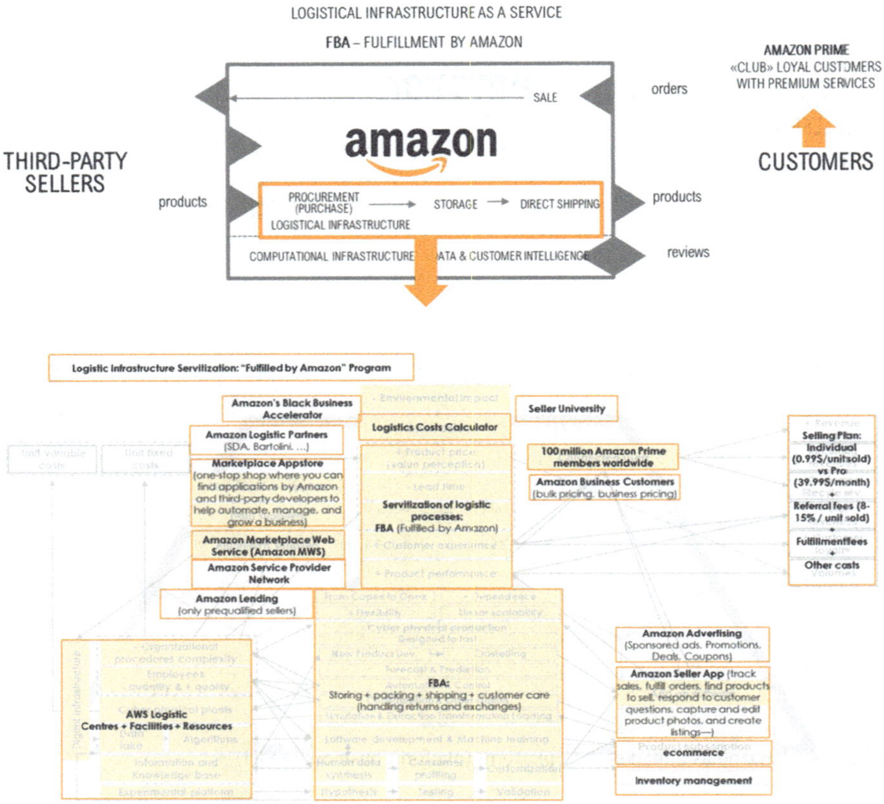

Fig. 4.5 Amazon as a service provider (data-driven and servitization business models): fulfilment by Amazon (FBA)

4.3 Amazon as a Service Provider (Data-Driven and Servitization Business Models)

Amazon has obtained great competitive advantages in embracing the potential of the as-a-service business model related to Industry 4.0 (Fig. 4.5).

Amazon has seized the opportunity to develop its as-a-service business model following the introduction of fulfilment by Amazon (FBA—shipped by Amazon), the service through which Bezos has "opened" its logistics platform allowing third-party sellers to store and ship their products directly from Amazon's fulfilment centres (FC), therefore leveraging partnerships with major logistics operators (e.g. SDA, Bartolini, etc. in Italy). Over time, third-party sellers have also been able to take advantage of Prime shipping by joining the FBA service, thus having access to a pool of over 100 million Amazon customers. Therefore, it is easy to see why this service has been so successful. In order to simplify and improve the customer experience with sellers, Amazon has made available an online calculator to define the costs of logistics services and has created a modular and flexible commercial offer model (e.g. monthly flat amount vs. per unit sold) to take into account different needs (e.g. small vs. large sales volumes).

Over time, in addition to logistic processes, Amazon has continued to develop its servitization business model in a structured way.

The Amazon Marketplace Appstore consists of the store of applications developed by Amazon and third parties and is available to enterprises in order to enable the automation, management and growth of their businesses. The Amazon Marketplace Web Service (Amazon MWS), now evolved into the Selling Partner API (application programming interface) (SP-API), has been made available to third-party developers and sales partners as a suite of API-based automation capabilities and authorization protocols for programmatic access to critical Amazon functions. The Amazon Service Provider Network, on the other hand, represents the network of service providers accredited by Amazon to support enterprises and sellers in developing their own business on Amazon.

Amazon has also developed advertising services for sellers to manage promotions, deals, coupons and sponsored ads on its e-commerce. The Amazon Seller app has been made available to enable sellers to better manage their business on Amazon, with features for monitoring and processing orders, answering customer questions and so forth.

In order to support enterprise business growth through the platform, Amazon has introduced the Amazon Lending suite of financial services.

Finally, it created the Amazon Seller University suite of training services and content to empower sellers and businesses to learn how to best leverage Amazon's capabilities.

But it is certainly with the introduction of the Amazon Web Services (AWS) that Amazon has established a successful business model paradigm, giving considerable impulse to the adoption and diffusion of cloud computing. The Amazon Web Services was initially created to solve precise internal needs of the group; later it was made available as a solution and technological infrastructure for external customers and still constitutes one of the fastest growing and most profitable services for the company.

In a similar fashion to LEGO bricks, AWS provides the technology fundamentals (building blocks) that enable developers and enterprises to generate and customize their own software application solutions. The information technology infrastructure becomes a service. The production of software applications becomes customized through the reconfiguration of these technology fundamentals. In this way, the market demand dynamically adapts to production; this is also favoured by the absence of investment costs in proprietary infrastructure (Fig. 4.6).

Following the introduction of the Amazon Web Services (AWS), Amazon has created a very profitable business model and has also taken a leading role in the emergence of cloud computing technology. The paradigm shift consists of having moved from the infrastructure as a product to the infrastructure as a service, thus

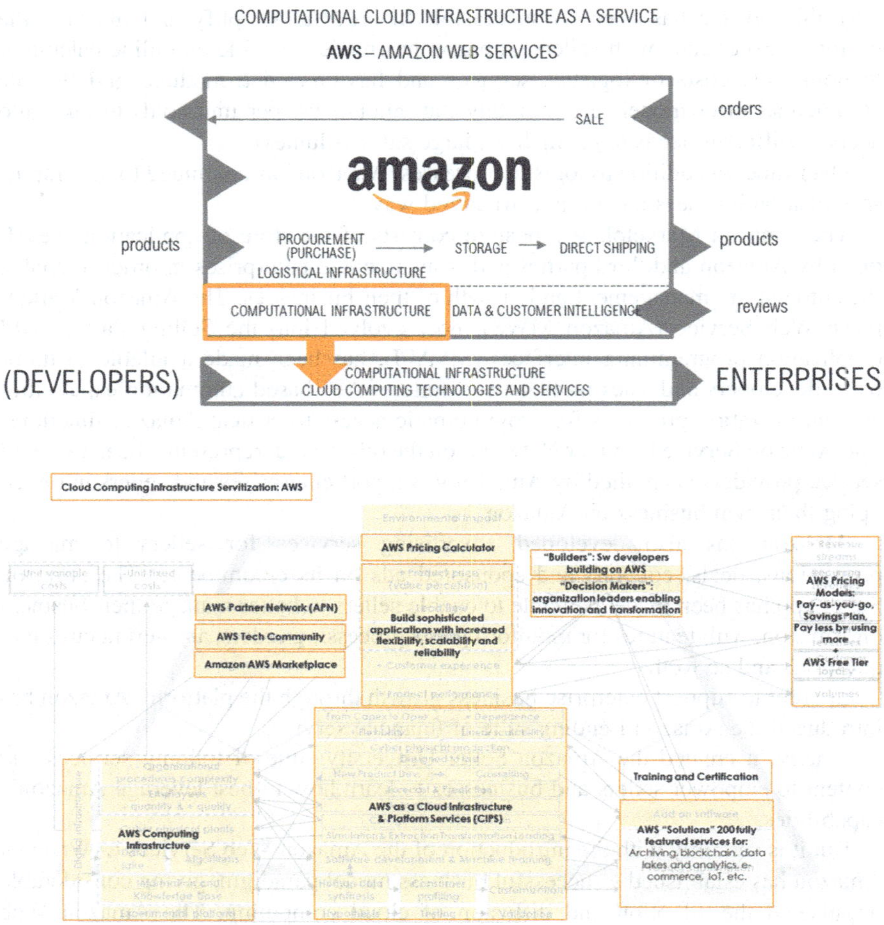

Fig. 4.6 Amazon as a service provider (data-driven and servitization business models): Amazon Web Services

offering a pay-per-use pricing model. The IT infrastructure is no longer acquired as an asset, entailing high fixed costs in terms of CAPEX (capital expenditure); it instead becomes a cloud service, that is, a variable cost linked to its usage in terms of OPEX (operating expense).

This has allowed start-ups, Newco's and companies of any size to start ICT projects, digital transformation or simply software development processes, relying on more sustainable investments as they are associated with balanced cash flows.

Moreover, being a technological infrastructure provided as-a-service and integrated into the developed application allows to build strong and long-term relationships with customers.

Amazon AWS, therefore, represents the base, a set of solutions (storage space, database, computing power, flexible payments, messaging, IoT, etc.), referred to as technology fundamentals that are rearranged and assembled by developers around the world for the purpose of designing "sophisticated and scalable" applications and services. Amazon has thus become, thanks to the introduction of AWS, the technology partner for the successful development of a digital business ecosystem based on cloud computing.

Companies like Amazon, which was born with strong commercial imprinting and sales orientation, have never underestimated the importance of collecting data and in particular customer data. Since its inception, Amazon learnt the potential of digital transformation, establishing itself as an e-commerce book shop has led to digital data tracking. A conversation with publisher and web technology expert Tim O'Reilly made Bezos realize the potential and true value of data. He realized that he possessed an immense amount of data, representing the core driver of competition and innovation for Amazon as well as for further companies interested in having access to data. From that moment, Amazon became an "algorithmic enterprise", that is, an organization capable of nurturing, generating and managing its business through data, involving customers, partners and additional stakeholders.

Data enable Amazon to identify customer wishes and preferences, which products to push as they attract customers; which product positioning, promotion and pricing strategies are more effective, also considering what its own "partners" are doing while selling their products on Amazon; etc. Data allow Amazon to manage its business in an "intelligent" way (in the literal sense), that is, to understand what is happening in real time (when) and to discover the current trends (what), reaching microsegmentation and examining the individual user (who) and his/her own behaviour (how) and especially his/her own extrinsic and intrinsic motivations and consumption choices (why).

For these reasons, Amazon's data-driven business model pursues both direct and indirect data monetization aimed at offering a unique "smart" customer experience. Part of Amazon's current competitive advantage derives from indirect monetization, which allows it to work on its downstream value proposition with respect to customers and on the upstream one and with respect to suppliers and partners, therefore introducing new services, products and categories within its (infinite) assortment, increasing the efficiency and productivity of its processes (Fig. 4.7).

Amazon's enterprises and business users have been particularly interested in being able to capture sales statistics and performance in reference to competing products, insights into market sectors, trends and tendencies. Some data is available on the Amazon sales rank and bestsellers section. The Movers and Shakers webpage highlights the products that are selling the most at that moment. Customers themselves are interested and influenced at the same time by these data: on the one hand, data represent a means for guaranteeing transparency and providing information to its customers; on the other hand, data constitute a marketing initiative to convince and reassure but also influence customer purchasing decisions. As for books, data help book authors understand whether their publications are appreciated, an aspect that could lead some of them to check the rankings.

Fig. 4.7 Amazon as a service provider (data-driven and servitization business models): Amazon as an algorithmic data-driven business

The first major innovation for Amazon coincides with a form of direct monetization granted to other websites: the Associates Program introduced in the summer of 1996 proved so successful that some say it gave impetus to the entire affiliate marketing industry. Amazon paid an 8% commission on book sales to referring websites. Once Amazon affiliate users send customers to Amazon via a link on their website, they obtain a percentage of the sale of the goods purchased within the next 24 h. Therefore, it represented the first example of collaboration (under an ecosystem logic perspective) between stakeholders and Amazon itself.

In terms of direct data monetization, Amazon has developed its own demand-side platform (DSP) (https://advertising.amazon.com/) to allow third parties (companies, agencies, etc.) to publish targeted advertising on Amazon.com or other websites, intended for users characterized by a certain set of traits and whose needs match with the products or services being advertised on the web advertisement.

As far as Amazon is concerned, therefore, the user is the product of the core dimension of its business model: and if it is untrue that "if something is free, you're the product" (Facebook and Google), considering Amazon's business model, we can paraphrase the statement above in "whether something is free or not, you're (also) the product".

Amazon's recent move towards artificial intelligence opens further opportunities associated with direct monetization, as artificial intelligence is made available via APIs for the purpose of developing third-party programmes and interacting with the data available.

Finally, the creation of data is also due to customer engagement. This solution has been pursued by Amazon since its inception, as Amazon has made it possible to review products (books) purchased via its e-commerce. Nowadays, reviews are still an important feature on Amazon, even though some sellers undermine the authenticity, transparency and integrity of this solution. Also, in order to overcome this potential criticality, Amazon is currently investing in the Amazon Vine programme consisting of an invitation-only programme, where customers (based on parametric data) are asked to review products in a qualified manner and without apparent conflicts of interest with the producers/sellers.

A further form of user collaboration is represented by the Amazon Build It programme, allowing customers to vote through pre-orders for which products they would like to develop and market (crowdfunding project/similar to Kickstarter).

4.4 Amazon as a Phygital Product/Store Platform (Data-Driven and Platform Business Models)

At the core of Amazon's strategy and business model remains (an obsession for) the customer experience. Amazon has been a digital-first company since the beginning, introducing consumers to the concepts of digital transformation and e-commerce. In this sense, Bezos seized the potential of getting to know all its customers, which is something that "traditional" retailers never had the opportunity to do [1].

The outstanding customer service and experience must be realized: the real pay-off for Amazon and its founder occurs while helping customers make purchase decisions. Amazon's mantra for its founder is to "raise the bar across industries, and around the world, for what it means to be customer focused" [1].

This means being an un-store that is rejecting the established rules and customs of traditional business and only caring about what is best for your customers. Amazon was meant to be digital-first since the beginning, and for this reason, it was not bound by what its predecessors had developed.

Amazon promises to allow customers to navigate through its website and purchase both what they want and what they did not imagine they would want. The other promise is to be committed to customer satisfaction [1], offering them low prices every day, providing them with the essential tools to check that the prices applied are the lowest (third-party seller prices, presence of pricing bots and sites where infographics are illustrated with price trends of individual items over time) and deploying its competitive supremacy for the benefit of customers.

Amazon also allows to fulfil the demand for instant-gratification and strives to ensure that such gratification is delivered to customers in the shortest time possible.

In doing so, Amazon empowers customers to influence their shopping habits and behaviours, sometimes changing their thinking patterns. For example, by personalizing the contents of its site on the basis of historical purchases and customer interests, it has designed the offer of its store around the customers themselves; with their recommendations and suggestions, it "exploited" the power of its customers to create trust and be useful to other customers; with Free Saver shipping, it has managed to get its customers to place larger orders and in different product categories; and with just one click, it pushed them to trust the platform and spend more, as well as with prime which has also contributed to building customer loyalty by promoting a club subscription for fast and flat rate shipping.

In this regard, the intuition related to the solution technology one click, which appears so simple but at the same time so successful and constituting an exclusive Amazon patent, contributed to reducing to a single click the so-called friction (all its competitors require two at least). As already mentioned, the web portal, smart apps and products contribute to transferring data from users to Amazon and vice versa, representing the digital and physical touchpoints of a customer experience that must be managed seamlessly in order to be effective and differentiating as Bezos learnt from Jobs' experience in Apple (Fig. 4.8).

Amazon has embraced digital transformation to develop its own business model based on the provision of customer-oriented services. In this case, it is more appropriate to refer to it as a hardware add-on rather than a software add-on business model, since the devices introduced by Amazon present a dual nature, digital-first and service-first, that is, they have been designed and created to promote Amazon services first instead of products to which a service component has been added.

Therefore, they represent devices whose value consists of the systemic integration of the product and service components. A recent example of a hardware add-on service is represented by the Dash button (now discontinued), a device that allowed the user through a simple "physical" click to reorder directly on Amazon a specific item, previously associated to the specific device through a simple configuration procedure. It was still a reorder that starts from the user, but certainly in the future, everything can be transformed into a more "predictive" service as it is algorithmic (as is already done, for example, with regard to diapers, relieving new parents from the burden of having to reorder them). The same can be implemented, for example, for the delivery of fruit and vegetables through Amazon fresh at home, activating a subscription-based service. Today, several product categories on Amazon can be purchased in the "recurring" mode, setting a quantity and a period of time that automatically schedules the repurchase and home delivery of a particular product.

In the case of Amazon, the value of its services is not so much linked to the ability to monitor the product, as is the case instead for a servitization business model in the B2B or B2C sphere that is product based: in fact, the Amazon product serves to monitor the services and acts as an access interface towards them. This is true for Echo-Alexa devices and Fire TV and in part for e-readers and home security, which increasingly assume the function of a constantly connected and updated instrumental platform that combines sensor, connectivity and data generation technologies with access to services and applications that interact with the user and enrich and qualify the overall experience.

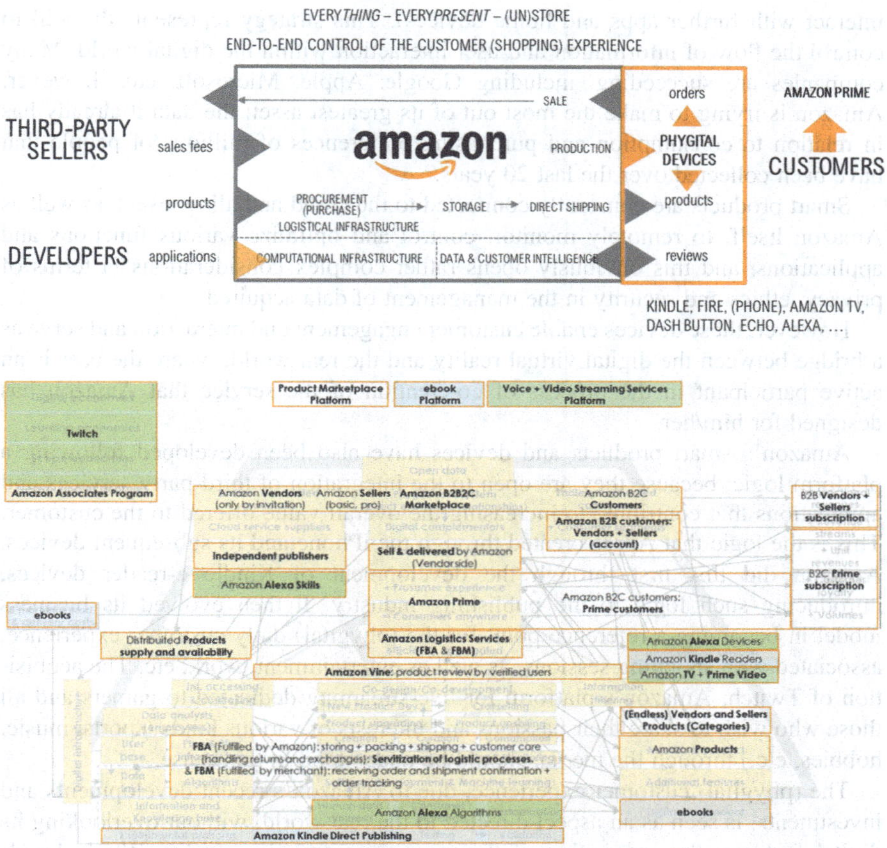

Fig. 4.8 Amazon as a phygital product/store platform (data-driven and platform business models): the phygital devices

Once these services meet customer needs, they contribute to building trust, as well as to creating a lock-in strategy with the Amazon ecosystem.

Amazon's main product in a broad sense currently remains the web portal (Web and app), as the digital place where interactions with the user take place and the primary context for collecting data from customers. In the meantime, Amazon has understood that the interfaces and touchpoints, aimed at a better and more pervasive customer (shopping) experience, could (and should) be diversified and innovated thanks to the availability of new technologies. For this reason, it designed and realized real Amazon B2C smart products, such as Amazon Echo and the virtual assistant Alexa, and physical devices destined to become the main channel to collect data from customers and transmit them through the cloud. At the same time, thanks to the cloud and the artificial intelligence system, these devices allow to return value to the customer (e.g. answer questions, make proposals based on data collected, interact with further apps and home devices). This strategy represents the will to control the flow of information and user interaction within the digital world. Many companies are succeeding, including Google, Apple, Microsoft, etc.; however, Amazon is trying to make the most out of its greatest asset: the data it already has in relation to consumption and purchasing preferences of millions of people that have been collected over the last 20 years.

Smart products are constantly connected to the cloud and allow users, as well as Amazon itself, to remotely monitor, control and optimize various functions and applications, and this obviously opens rather complex considerations in terms of privacy, ethics and security in the management of data acquired.

However, these devices enable customer engagement and interaction and serve as a bridge between the digital virtual reality and the real world, where the user is an active participant in the process of co-creation of the service that Amazon has designed for him/her.

Amazon's smart products and devices have also been developed following a platform logic, because they are open to the integration of third-party services and applications that contribute to increasing the overall value offered to the customer. This is the logic that Apple created through the iPhone and its subsequent devices. Amazon did this first through the development of Kindle e-reader devices, introducing such logic to the publishing industry. It then evolved its business model in becoming a reference point for the (phygital) daily customer experience, associated with shopping sessions, as well as entertainment, work, etc. The acquisition of Twitch, Amazon's platform for live-streaming dedicated to gamers and all those who wish to share their passions and interests of various kinds (sports, music, hobbies, etc.) through the media, should be understood in this sense.

The (phygital) customer experience, due to Amazon's recent developments and investments, is seen as an aspect directed to the real world, without overlooking its digital features: the acquisition of the premium—organic—quality Whole Foods stores, the introduction of the automated Amazon Go stores, the opening of Pick-up points and recently the return to the "physical" Amazon Bookstore stores, including direct home delivery to cover even the last mile (Fig. 4.9).

In conclusion, Bezos has understood the strategic meaning [4] behind its everything-un-store and has succeeded in changing and redefining the rules of the game. Therefore, Amazon represents a paradigmatic case of strategic innovation [5] made possible, as we have discussed, also thanks to the integration and systemic development of the different business models characterizing Industry 4.0. The so-called world amazonification [5] continues, seemingly in an unstoppable way, and is perhaps only at its beginning.

4.5 Identifying the Tools and Techniques of Assessment

Business model innovation requires organizations, willing to disrupt their traditional business models, to develop digital strategies as well as competitive strategies in order to drive innovation and business growth, thereby transforming industries. Both strategies result intertwined and therefore must be consistent with one another.

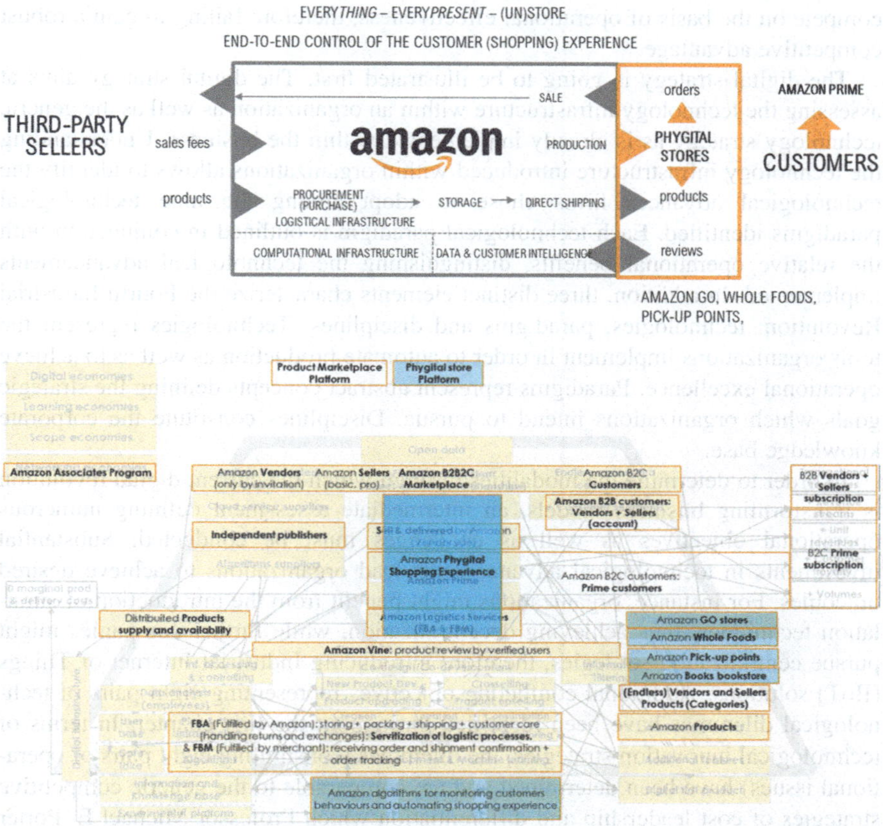

Fig. 4.9 Amazon as a phygital product/store platform (data-driven and platform business models): the phygital stores

Nonetheless, management fails to distinguish between strategies and operational effectiveness, thus implementing management tools in order to enhance operational effectiveness, while failing to provide organizations with sustainable profitability. As Professor Michael E. Porter declares, strategies must not be confused with operational effectiveness. Although strategy and operational effectiveness appear to be crucial to achieving superior performance, both options operate differently. The former describes the process of performing distinct organizational activities compared to rivals or of performing similar activities in a different manner in order to deliver greater value to customers or to generate comparable value at a lower cost. The latter refers to companies performing superior activities as opposed to competitors, thus achieving organizational effectiveness as well as excellence. Furthermore, organizations can outperform rivals by establishing a difference to be preserved. Nevertheless, Professor Michael E. Porter argues that most organizations compete on the basis of operational effectiveness, therefore failing to gain a robust competitive advantage.

The digital strategy is going to be illustrated first. The digital strategy aims at assessing the technology infrastructure within an organization as well as the generic technology strategy as is already implemented within the business. Understanding the technology infrastructure introduced within organizations allows to identify the technological advances firms chose to adopt among the nine technological paradigms identified. Each technological paradigm is outlined in conjunction with the relative operational benefits, distinguishing the technological advancements implemented. In addition, three distinct elements characterize the Fourth Industrial Revolution: technologies, paradigms and disciplines. Technologies represent the tools organizations implement in order to automate production as well as to achieve operational excellence. Paradigms represent abstract concepts defining the strategic goals which organizations intend to pursue. Disciplines constitute the corporate knowledge base.

In order to determine the modalities through which the current digital revolution is transforming business models, an intermediate assessment defining numerous operational objectives as well as challenges must be conducted. Substantial investments in technological advancements lead organizations to achieve desired outcomes. For instance, organizations might benefit from the introduction of simulation techniques, thus achieving decentralization, while further companies might pursue centralization strategies, therefore introducing Industrial Internet of Things (IIoT) solutions. Additional conflicting objectives representing eight pairs of technological dilemmas have been identified throughout the first chapter. In terms of technological innovation strategies, the dilemmas (constituting eight pairs of operational issues) have been determined and are comparable to the generic competitive strategies of cost leadership and differentiation which Professor Michael E. Porter has defined as distinct approaches enabling organizations to achieve sustainable competitive advantage. Furthermore, literature reviews have been conducted in order to establish the operational benefits related to the implementation of the principal technological paradigms within the distinct building blocks, constituting the business model canvas. The results have been summarized within a table illustrating the effects of the introduction of technological advances in manufacturing processes. The example relating to digital additive manufacturing is reported below. Technological paradigms comprise additive manufacturing, whose implementation contributes to reducing the number of suppliers as well as minimizing costs and lead times. Additive manufacturing also implies the introduction of three-dimensional printing equipment, thus avoiding the need to manufacture individual complex components and printing the required object in a single run. Each operational benefit identified is associated with either the achievement or the non-fulfilment of its corresponding operational objective, based on a theoretical perspective (Table 4.1).

In order to determine the generic digital strategies pursued by organizations, consultants assign a score ranging from −5 to +5 to each dilemma depending on the strategic orientation of each element. The overall distribution of the scores

Table 4.1 Digital paradigms: digital manufacturing

Building block	Operational benefits	Decentralization	Centralization	Externalization	Internalization	Interoperability	Ownership	Usability	Security	Prediction	Reactivity	Augmentation	Automation	Flexibility	Efficiency	Complexification	Simplification	Value proposition
Suppliers	Reduces the number of suppliers in particular of those specialized in the production of complex parts (self-production of components)		✓		✓		✓	(✓)	(✓)		✓	(✓)	(✓)	✓			✓	Price, availability, accessibility, rapidity
Resources	Reduces the use of unskilled labor and therefore the convenience of low-cost relocation strategies	(✓)	(✓)		✓	(✓)	(✓)	(✓)	(✓)		✓		✓	✓		✓		Reliability, material quality, performance
	Reduces the optimal size of production sites (reduces economies of scale)	(✓)	(✓)	✓		(✓)	(✓)	(✓)	(✓)		✓	(✓)	(✓)	✓			✓	Availability, accessibility, rapidity
	Reduces the number of items in stock (elimination of components for products, self-production of some components)	(✓)	(✓)	(✓)	(✓)	(✓)	(✓)	(✓)	(✓)	(✓)	(✓)	(✓)	(✓)		✓		✓	Price, availability, accessibility, rapidity
Internal Processes	Reduces the time and costs (including customs) of procurement (self-production of components)	(✓)	(✓)		✓	(✓)	(✓)	(✓)	(✓)		✓	(✓)	(✓)		✓		✓	Price, availability, accessibility, rapidity
	Reduces product development time and costs (time to market and learning economies)	(✓)	(✓)	(✓)	(✓)	(✓)	(✓)	(✓)	(✓)		✓	(✓)	(✓)	✓		(✓)	(✓)	Newness
	Reduces prototyping times and costs by digitizing the product realization process	(✓)	(✓)	(✓)	(✓)	(✓)	(✓)	(✓)	(✓)		✓	(✓)	(✓)	✓			✓	Reliability, newness
	Reduces the time and costs for the production of small batches (lead time eliminating physical setup and reducing learning economies)	(✓)	(✓)	(✓)	(✓)	(✓)	(✓)	(✓)	(✓)		✓	(✓)	(✓)	✓			✓	Price, availability, accessibility, rapidity, product range and customization
	Reduces overproduction costs by adapting to market dynamics in real time	(✓)	(✓)	(✓)	(✓)	(✓)	(✓)	(✓)	(✓)		✓	(✓)	(✓)	✓			✓	Price, availability, accessibility, rapidity
	Increases freedom in product design by fostering individual creativity and the birth of new ideas	(✓)	(✓)	(✓)	(✓)	(✓)	(✓)	(✓)	(✓)		✓	✓		✓		✓		Newness, immaterial quality, life (work) style

contributes to indicating the generic digital strategy introduced. Organizational choices can lead to the following four distinct digital strategies (Fig. 4.10):

1. Open and Freedom Strategy. The open and freedom strategy encourages organizations to pursue openness and freedom in order to enable the achievement of an emerging balance.
2. Neutral Strategy. The neutral strategy represents a strategic choice enabling businesses not to pursue particular digital strategies.
3. Zigzag Strategy. The zigzag strategy consists of a comprehensive as well as an innovative strategic option, figuring as a combination between the open and freedom strategy and the closed and control strategy. Furthermore, the present alternative constitutes the optimal choice for organizations as well as the foundation for the development of an advanced competitive strategy.
4. Closed and Control Strategy. The closed and control strategy encourages enterprises to pursue closure and control in order to enable companies to achieve a certain balance.

Subsequently, the alignment between the generic as-is digital strategy and the generic to-be digital strategy must be verified. Nonetheless, the methodology illustrated herein has yet to be tested.

4.5.1 The Operational Fit Within Industry 4.0: Digital Strategy

Consultants are required to guide chief digital officers (often shortened as CDO) as well as chief innovation officers (often abbreviated to CINO) in the selection of the optimal digital strategy according to the organizational aspirations, thus supporting management in developing a global vision. Firms are required to complete a survey or a questionnaire, therefore revealing the corporate technology infrastructure as well as the generic digital strategy implemented. Consequently, the results are validated utilizing a set of theoretical frameworks, in order to compare the actual digital strategy with the expected outcomes. Therefore, data collection serves to determine the corporate digital strategy, among the four generic digital strategies previously illustrated. On the other hand, data analysis serves to formulate an action plan in order to achieve the alignment between the generic digital strategy as-is and the generic digital strategy to-be.

Surveys aim at investigating the generic digital strategies implemented. In this regard, the management perspective figures as essential in establishing the digital strategies pursued. Furthermore, management must determine the technological paradigms introduced and the effects each technological advancement produced within the distinct building blocks, constituting the business model canvas. Finally, corporate management must highlight the operational benefits derived, resulting from the implementation of the technological paradigms identified.

Data collected is then examined for the purpose of enabling consultants to evaluate the alignment between the results that emerged upon completion of the

Fig. 4.10 Generic digital strategies

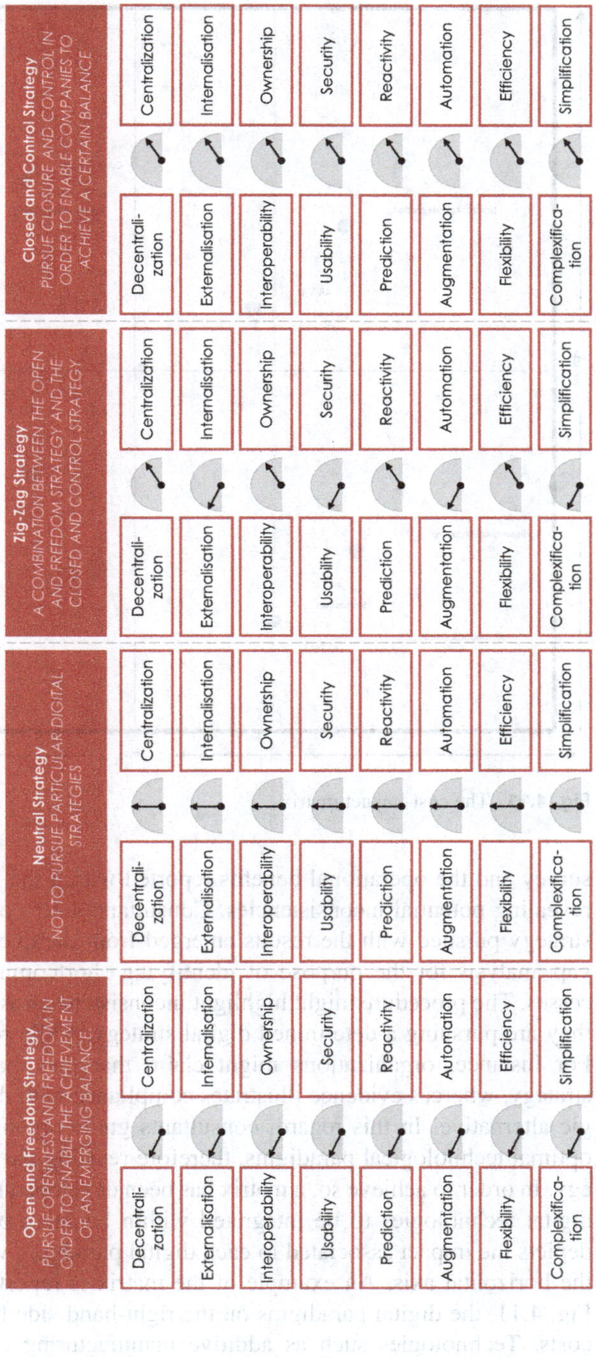

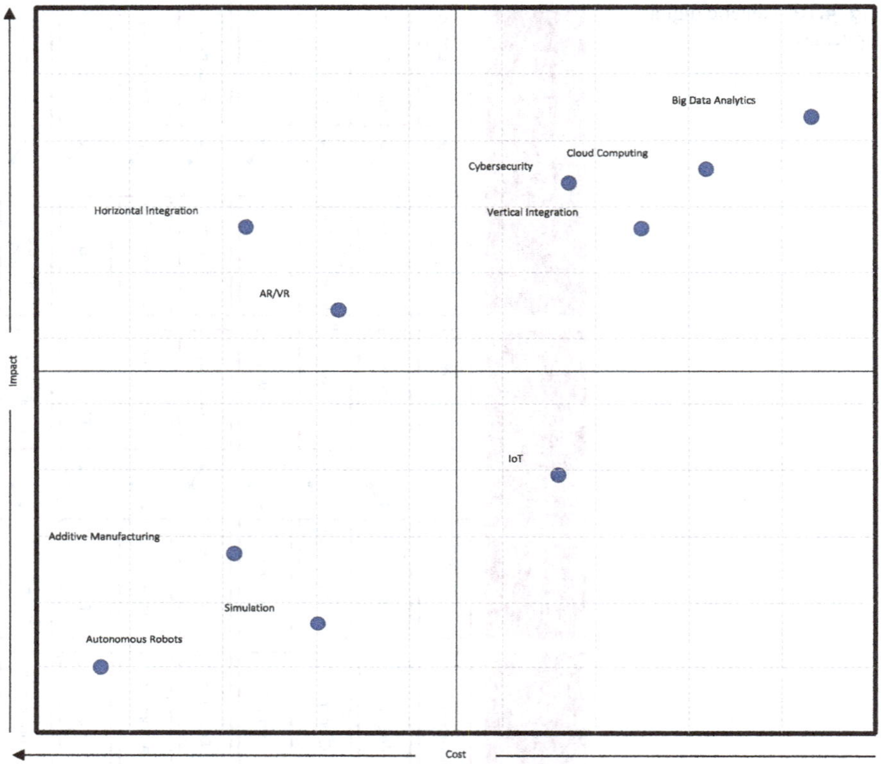

Fig. 4.11 The cost-impact matrix

survey and the operational benefits reported within the theoretical framework, thus revealing potential inconsistencies. Consultants later compare the corporate digital strategy pursued with the results emerged from the questionnaire, thus conducting gap analysis for the purpose of identifying shortcomings affecting business processes. The procedure might highlight inconsistencies as often organizations believe they are pursuing a determined digital strategy, while results suggest the opposite. For instance, organizations might claim that they have implemented a zigzag strategy, whereas evidence illustrates compliance with the open and freedom strategic alternative. In this regard, consultants guide organizations in determining the optimal technological paradigms, therefore revising the digital organizational strategy. In order to achieve so, a matrix has been developed to facilitate the choice of the digital technologies to be integrated within business processes. The vertical axis depicts the impact associated to each digital paradigm, while costs are measured on the horizontal axis. An example of the matrix is reported below. As illustrated in Fig. 4.11, the digital paradigms on the right-hand side have a high impact and low costs. Technologies such as additive manufacturing and simulation present low impacts and high costs.

Consultants utilize the matrix for the purpose of supporting companies in determining the choice of the technological paradigms to implement within business processes based on specific organizational goals. Once the appropriate digital paradigms have been identified, the assessment process then requires listing and prioritizing use cases and exploring existing solutions as well as performing a make-or-buy analysis, thus establishing whether a particular work can be accomplished inside the firm or should be purchased from external sources. Depending on the result obtained from the analysis, the solution can be either created from scratch inside the organization or co-created or purchased from an outside firm. Solutions must then be tested and integrated.

During the final stage of the assessment (which is discussed in the following paragraphs), consultants propose the following three distinct options to management:

1. Add Technology. Organizations can choose to introduce an additional digital paradigm.
2. Remove Technology. Consultants can encourage management to quit investing in determined technological paradigms which do not contribute to the achievement of the desired digital strategy.
3. Change Implementation. Consultants can support companies in designing innovative technologies for the purpose of achieving multiple operational benefits.

4.5.2 The Competitive Fit Within Industry 4.0: Competitive Strategy

The corporate digital strategy must be developed following the organizational competitive strategy. Nonetheless, often management responsible for formulating the digital strategy does not entirely understand the business initiatives characterizing the competitive strategy. Therefore, consultants must communicate with chief digital officers (CDO) in order to formulate the desired digital strategies.

In order to achieve complete alignment, consultants must assess the competitive strategy as-is. The competitive strategy formulation follows a similar procedure as the methodologies previously described for the purpose of evaluating the corporate digital strategy implemented. The process relating to the identification of the competitive strategy introduced with the organization requires consultants to interview management in order to highlight the effects that business processes produce within the building blocks, constituting the business model canvas. For instance, management should identify the most critical and distinctive organizational resources representing a source of competitive advantage.

Recently, consultants have succeeded in interviewing management, thus collecting valuable data as well as information in order to help organizations formulate a successful comprehensive strategy. Interviews have contributed to highlighting interesting results. In particular, management identified manufacturing equipment and real estate assets among the most valuable resources. The results are illustrated in Table 4.2.

The disclosed findings have been then reported within the building blocks characterizing the business model canvas, hence gaining insights into the organization's business model (Fig. 4.12).

The following step characterizing the assessment consists of transferring the results highlighted within the theoretical categories onto the competitive fit matrix for the purpose of determining the organizational strategic fit. Competitive fit is achieved once organizations present alignment along all building blocks. According to the distribution of the elements constituting the competitive fit matrix, it is possible to identify the following four major competitive strategies:

1. Operational Excellence. Alignment towards the right-hand side of the matrix suggests organizations are pursuing operational excellence, thus manufacturing standardized goods and services and aiming to accomplish cost leadership.
2. Product Leadership. Alignment towards the right-hand side of the matrix also suggests businesses intend to pursue product leadership, therefore generating innovative and high added value goods and services as well as achieving premium prices.
3. Ecosystem Leadership. Alignment towards the left-hand side of the competitive fit matrix suggests companies are oriented towards ecosystem leadership, therefore orchestrating resources and managing transactions within platforms.
4. Customer Intimacy. Alignment towards the left-hand side of the strategy fit matrix also suggests firms' aim at pursuing customer intimacy, therefore

Table 4.2 Competitive fit: building block identification

Complete alignment *(21 – 31 of the respondents)* ***Partial alignment*** *(11 – 20 of the respondents)* ***Low alignment*** *(0 – 10 of the respondents)*

THEORETICAL CATEGORIES	MANAGEMENT STATEMENTS				
Tangible capital 25	Physical	22	Machines and equipment	12	
			Real estate asssets	17	Production plants
	Financial	7	Assets	1	
			Liquidity	1	
Structural capital 21	Intellectual property	13	Patents	1	
			Authorizations and permissions	1	
			Information database	1	
	Organizational	10	Organizational procedures	8	
			ICT	5	
Human capital 21	Hard skills	14	Education	5	
			Experience	8	
	Soft skills	9	Individual	3	
			Collectives	7	
Relational capital 12	Operative	11	Suppliers relations	3	
			Customers relations	4	
			Image and reputation	9	Product brands
Environmental capital 1	Natural	1	The HQ and its surroundings		

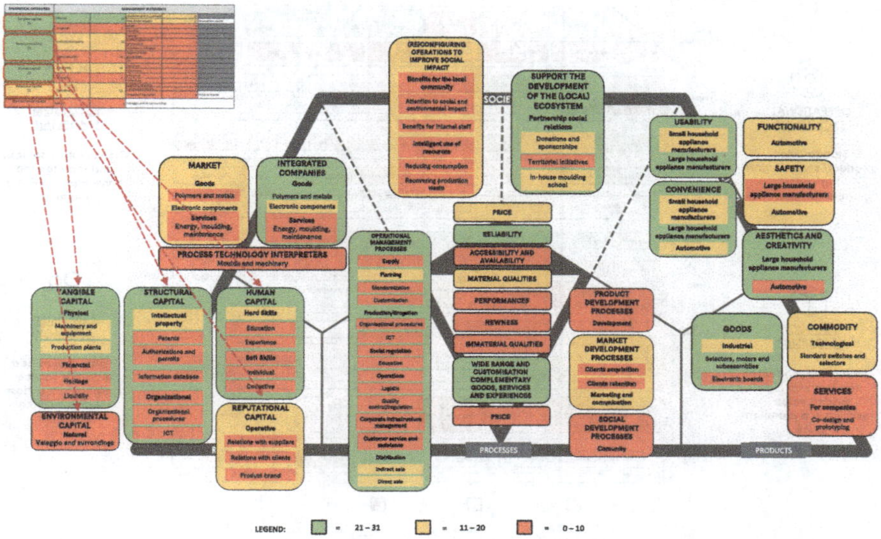

Fig. 4.12 Competitive fit: business model definition

manufacturing highly customized products as well as services as a result of a closer relationship with customers.

Nevertheless, Fig. 4.13 presents a rather disparate distribution pattern; thus, the organizational strategic fit did not appear as solid as management would have expected.

Following the distribution pattern obtained, management can decide whether to enhance the value proposition, thus potentially improving involvement and sharing. The choice to enhance the value proposition in this case requires organizations to improve all of the further elements positioned below.

Considering the example reported above, the business aimed at differentiating itself from competitors, thus leveraging ongoing customer relationships and fulfilling evolving consumer demand. Nonetheless, customer intimacy did not appear sufficiently valued. Therefore, the company managed to enhance certain strategic areas through the introduction of innovative technologies, the distribution of a unique range of customer services as well as consumer experiences and the market redefinition in order to address consumer needs.

The ultimate goal represents achieving strategic convergence, thus verifying whether the generic digital strategy developed enables the organization to achieve a strategic fit. Therefore, enterprises must align the organizational competitive strategy to the digital corporate strategy previously formulated. Furthermore, management is required to conduct the operational fit analysis, thus determining the digital infrastructure characterizing the organization in order to align the actual digital strategy to the expected outcomes, thus sustaining superior performance

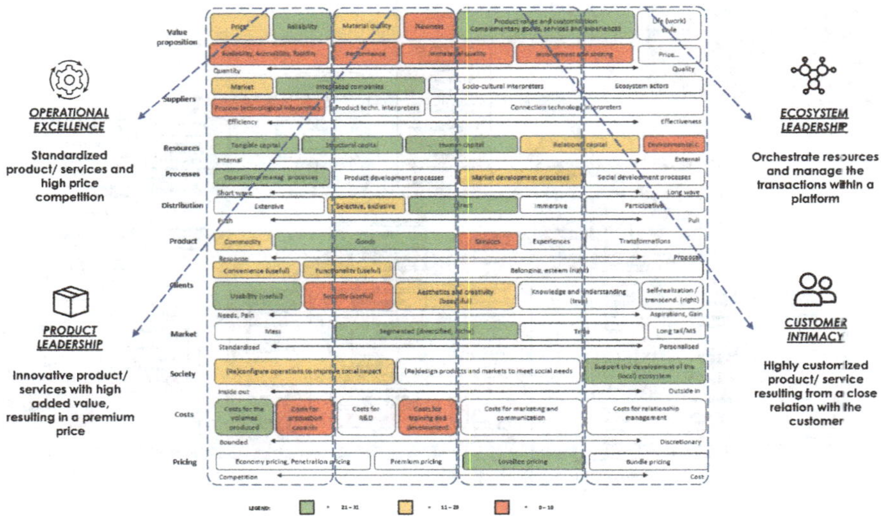

Fig. 4.13 Competitive fit: competitive fit analysis

and reaching corporate goals as well as targets. In addition, consultants must verify that alignment exists between the operational benefits that emerged following the interviews and the results illustrated within the theoretical framework. Following the theoretical perspective, each operational benefit can be associated with particular impacts characterizing the value proposition as well as further building blocks, constituting the business model canvas. For instance, the introduction of additive manufacturing and the consequent reduction in the number of suppliers contribute to increasing efficiencies along the supply chain, thus enhancing the product's reliability. Moreover, it is possible to depict the impact of the digital strategy on the value proposition. As a result, the digital organizational strategy formerly identified can contribute to enhancing product quality as well as to reducing product price. Subsequently, consultants are able to revise the digital strategy, comparing the future digital strategy with the competitive corporate strategy. Although both the digital and the competitive strategies provide distinct manners to enhance the value proposition as well as further dimensions, organizations are allowed to choose among three alternatives in order to reach the desired effects on the value proposition, comprising implementing further technologies, removing technological solutions and readapting technological paradigms to organizational processes. Furthermore, consultants guide organizations in determining the digital paradigms contributing to enhancing the elements constituting the value proposition that have yet to be achieved, thus enabling the organization to achieve strategic convergence. Companies utilize the aforementioned matrix in order to determine the impacts as well as the costs associated with the choice of a particular digital paradigm needed to fulfil organizational strategic goals.

Currently, organizations are embracing digital transformation for the purpose of enhancing operational excellence, minimizing costs and achieving corporate sustainable development. Furthermore, the digital revolution is encouraging companies to introduce multiple technological advances in order to automate as well as to digitize manufacturing processes, thus contributing to generating enormous amounts of data. In addition, data enables the creation of innovative digital-first business models. Digital-based business models constitute business models based on data gathered as a result of the digitization of production processes. The methodology proposed within the present research ideally represents the procedures organizations must follow in order to implement digital business models and seize the numerous benefits characterizing the introduction of technological advancements within production processes. Companies disrupting traditional business models are going to generate greater customer value as well as gain significant competitive advantages.

Business model innovation is increasingly gaining prominence in practice as well as in academic literature. The unpredictable changes in digital technologies are contributing to questioning the robustness of an organization's current business model. Nonetheless, the theoretical procedures illustrated herein have been partially tested within organizations. Currently, Professor Carlo Bagnoli and fellow researchers are exerting increasing efforts in order to formulate a further practical approach. The strategy is going to support companies in strengthening their existing business models as well as implementing innovative digital-first business models, for the purpose of enabling businesses to leverage the multiple opportunities offered by the digital transformation.

References

1. Stone, B. (2014). *Vendere tutto. Jeff Bezos e l'era di Amazon*. Hoepli.
2. Forbes. (2021). *Amazon non è immortale. E il declino potrebbe essere già cominciato*. https://forbes.it/2021/10/04/amazon-non-immortale-potrebbe-avere-gia-iniziato-declino/
3. Bishop, T. (2013). *Bezos: 3D printing "exciting" but not disruptive for Amazon in short term*. GeekWire.
4. Battistella, C., Biotto, G., & De Toni, A. F. (2021). From design driven innovation to meaning strategy. *Management Decision, 50*(4), 718–743.
5. Bagnoli, C., et al. (2018). *Business Model 4.0. I modelli di business vincenti per le imprese italiane nella quarta rivoluzione industriale*. Edizioni Ca' Foscari.

5 Conclusions

The aim of this work is to analyse in depth the advent of the Fourth Industrial Revolution and how the technological advancements can be implemented into a company's business model. The first chapter delves into the theory of strategic innovation, the sources of innovation, and the frameworks necessary in order to conduct strategic analysis. The first chapter also describes the Fourth Industrial Revolution, defining its features and attempting to classify its underlying technologies according to a few known frameworks. The chapter concludes by citing the main operational drivers of the new industrial revolution. The second chapter analyses each of the nine enabling technologies in detail, identifying the main advantages in terms of operational benefits. The tables show more systematically the intersection among technologies, operational benefits and operational drivers and how they are connected with each other. In the third chapter, the results previously described are evaluated according to a strategic vision, presenting the new business models characterizing Industry 4.0 and proposing a final comprehensive framework. The research showed that digital-first business models may improve the ability of firms to respond effectively to rapid changes within the business environment. The fourth and last chapter illustrates a real-world business case study and suggests an assessment process guiding companies through the choice of the optimal digital paradigms and the development of digital as well as competitive strategies.

The biggest challenge consisted in adapting the new digital innovations to strategic innovation or business model innovation. Digital technologies not only improve processes but revolutionize the way a company creates, distributes and captures value. New data-driven business models are generated, thus breaking traditional barriers and opening the door to new profit opportunities. The Fourth Industrial Revolution is seen by many as a technological innovation, but few perceive it as a strategic innovation. The key is understanding how to interpret data and how to balance them with the customers, offering them the opportunity to actively participate within the company's value creation process.

C. Bagnoli et al., *Digital Business Models for Industry 4.0*, Future of Business and Finance, https://doi.org/10.1007/978-3-030-97284-4_5

Building relationships, both with customers and suppliers and with society, leads to creating long-lasting and stable relationships. Companies therefore need relational and creative skills, and they must also be able to attract and create a new set of skills and cultivate talents and in other words hire talented employees who are able to utilize the new digital technologies and enormous amounts of data. Nonetheless in order to achieve this, companies must have the courage to change and abandon traditional models, encouraging involvement in the ecosystem characterized by interactions between new roles and technologies.

From the results obtained from the matrix in the third chapter and from the tables in the second chapter, it follows that a new source of innovation arises: *technology push innovation*. This innovation can take two approaches: a bottom-up and a top-down approach. In the case of bottom-up technology push innovation, it is possible to start from an initial assessment of the company and focus on some specific operational benefits that the same company wants to enhance, in order to then identify which operational drivers have been affected the most and the related technologies of Industry 4.0. Alternatively, following the top-down technology push approach, we could start from the operational drivers (e.g. decentralization, interoperability) or the technologies which the company is interested in implementing into its processes and then identify the individual operational benefits, that is, the main advantages that they could activate. This dual approach between technology and strategy is characterized by a high degree of flexible reasoning, which implies being able to generate the same result beginning from different starting points.

The true potential of innovation is captured when following the exploitation of new technological opportunities; traditional business models are transformed into new paradigms characterized by innovative value propositions. Industry 4.0 is not only represented by its technologies; nevertheless, as discussed in this work, it is a process characterized by the convergence between different technologies. The implementation of these digital technologies within the business is not sufficient to capture the true potential of digital innovation; however, it is necessary to combine the benefits of technologies in a strategic manner depending on the organization, in order to create new business opportunities.

5.1 Appendix

5.1.1 Research Methodology

The research methodology consists of a structured literature review, used to examine the literature and summarize the information contained as well as to deepen the data collected. The initial step of the structured literature review is represented by the definition of search engines, such as Scopus, Google Scholar and Business Source Complete. The bibliography was selected by searching by keywords, such as "Industry 4.0", "business model" and "digital transformation". Research was conducted examining documents devoted to the topic of Industry 4.0 and published between 2018 and 2021, thus updating the previous work carried out by Professor

Carlo Bagnoli, and it showed that there is a growing interest in the issue of Industry 4.0. The papers were then added to the QDAS (qualitative data analysis software) database NVivo and manually coded. The coding structure of NVivo is reported below. Once the coding phase had been performed, the data were analysed through the query function that allows to cross the nodes. The procedure was validated by Professor Maurizio Massaro from Ca' Foscari University of Venice.

Name	Files	References
A_Type of report	168	170
A 01_Consulting firm	36	36
A 01.01_Report	30	30
A 01.02_Slide	6	6
A 02_Other	68	70
A 02.01_Institutional report	18	18
A 02.02_Other reports	18	20
A 02.03_Slide	32	32
A 03_Scientific papers	64	64
B_Business model	168	3704
B 1_Production	114	1602
B 1 01_Mass customization	75	305
B 1 02_Smart manufacturing	99	981
B 1 03 Hub and spoke	24	68
B 1 04_General	38	119
B 1 05_Examples	39	129
B 2_Consumption	66	298
B 2 01_Direct data monetization	13	23
B 1 03_Licensing	13	20
B 2 02_Indirect data monetization	22	61
B 2 03_Enhanced smart product's functionalities	40	107
B 2 04_General	29	74
B 2 05_Examples	14	33
B 3_Exchange	87	479
B 3 01_Add-on hardware	31	75
B 3 02_Add-on software	18	36
B 3 03_Servitization or smart product	48	128
B 3 04_General	39	177
B 3 05_Examples	25	63
B 4_Distribution	91	684
B 4 01_Open innovation	24	179
B 4 02_The digital and physical customer experience	41	106
B 4 03_Broker and technology platforms	51	195
B 4 04_General	29	76
B 4 05_Examples	27	86
B 4 06_New partnership network	19	42
B 5 Other	75	609

(continued)

Name	Files	References
B 6_Topic not present	28	32
C_Technologies	168	7577
C 01_Additive manufacturing	61	671
C 1.01_Resources	30	83
C 1.02_Internal processes	38	154
C 1.03_External processes	15	25
C 1.04_Products	32	78
C 1.05_Customers	18	37
C 1.06_Suppliers	11	14
C 1.07_Society	4	6
C 1.08_Value proposition	0	0
C 1.10_Description	35	141
C 1.20_Economic-financial impact	16	46
C 1.30_Examples	22	87
C 02_Advanced manufacturing solutions	83	892
C 2.01_Resources	52	224
C 2.02_Internal processes	59	226
C 2.03_External processes	9	22
C 2.04_Products	27	40
C 2.05_Customers	10	12
C 2.06_Suppliers	7	9
C 2.07_Society	4	14
C 2.08_Value proposition	0	0
C 2.10_Description	44	207
C 2.20_Economic-financial impact	17	65
C 2.30_Examples	26	72
C 03_Augmented reality	52	211
C 3.01_Resources	34	63
C 3.02_Internal processes	32	61
C 3.03_External processes	5	7
C 3.04_Products	8	9
C 3.05_Customers	3	3
C 3.06_Suppliers	1	1
C 3.07_Society	0	0
C 3.08_Value proposition	2	2
C 3.10_Description	27	39
C 3.20_Economic-financial impact	2	3
C 3.30_Examples	16	23
C 04_Cloud computing	80	459
C 4.01_Resources	31	95
C 4.02_Internal processes	36	68
C 4.03_External processes	6	10
C 4.04_Products	11	23
C 4.05_Customers	12	18

(continued)

Name	Files	References
C 4.06_Suppliers	12	34
C 4.07_Society	3	19
C 4.08_Value proposition	4	5
C 4.10_Description	47	126
C 4.20_Economic-financial impact	6	18
C 4.30_Examples	14	41
C 05_Simulation	40	211
C 5.01_Resources	14	30
C 5.02_Internal processes	27	84
C 5.03_External processes	7	8
C 5.04_Products	8	12
C 5.05_Customers	4	7
C 5.06_Suppliers	2	9
C 5.07_Society	3	6
C 5.08_Value proposition	0	0
C 5.10_Description	15	33
C 5.20_Economic-financial impact	6	9
C 5.30_Examples	9	13
C 06_Industrial IoT	105	1546
C 6.01_Resources	70	304
C 6.02_Internal processes	77	355
C 6.03_External processes	28	55
C 6.04_Products	52	135
C 6.05_Customers	28	70
C 6.06_Suppliers	24	46
C 6.07_Society	13	23
C 6.08_Value proposition	8	29
C 6.10_Description	71	313
C 6.20_Economic-financial impact	20	69
C 6.30_Examples	36	145
C 07_Big data and analytics	111	1554
C 7.01_Resources	74	411
C 7.02_Internal processes	80	432
C 7.03_External processes	28	42
C 7.04_Products	50	126
C 7.05_Customers	41	97
C 7.06_Suppliers	28	46
C 7.07_Society	12	19
C 7.08_Value proposition	12	23
C 7.10_Description	68	224
C 7.20_Economic-financial impact	25	45
C 7.30_Examples	30	79
C 08_Cybersecurity	51	673
C 8.01_Resources	27	111

(continued)

Name	Files	References
C 8.02_Internal processes	28	195
C 8.03_External processes	3	3
C 8.04_Products	10	19
C 8.05_Customers	7	12
C 8.06_Suppliers	8	17
C 8.07_Society	7	30
C 8.08_Value proposition	1	1
C 8.10_Description	33	232
C 8.20_Economic-financial impact	9	31
C 8.30_Examples	8	22
C 09_Structural horizontal and vertical system integration	49	440
C 9.01_Resources	22	47
C 9.02_Internal processes	30	100
C 9.03_External processes	16	28
C 9.04_Products	15	21
C 9.05_Customers	25	57
C 9.06_Suppliers	23	60
C 9.07_Society	10	19
C 9.08_Value proposition	6	8
C 9.10_Description	24	79
C 9.20_Economic-financial impact	7	10
C 9.30_Examples	6	8
C 10_Other	105	896
C 11_Topic not present	22	24
D_Industry 4.0 (general)	168	14,893
D 0 Capgemini	107	5829
D 0.01 value chain (base)	105	4129
D 1.01_Resources	94	1376
D 1.02_Internal processes	99	1709
D 1.03_External processes	75	350
D 1.04_Products	91	694
D 0.02 value system (half)	88	1215
D 1.05_Customers	85	721
D 1.06_Suppliers	74	494
D 0.03 Ecosystem (top)	67	485
D 1.07_Society	67	485
D 1.01_Resources	113	1447
D 1.02_Internal processes	113	1740
D 1.03_External processes	81	366
D 1.04_Products	107	748
D 1.05_Customers	104	769
D 1.06_Suppliers	86	519
D 1.07_Society	79	513
D 1.08_Value proposition	35	120

(continued)

Name	Files	References
D 1.10_Description	68	367
D 1.20_Economic-financial impact	90	563
D 1.30_Examples	63	296
D 1.40_General	102	1529
D 1.50_Competitor	32	59
D 1.60_Topic not present	26	28
E_Roadmap (consulting process)	168	943
E 01_Process	9	34
E 02_Description	26	58
E 03_Specific consulting services	1	1
E 04_Customer needs	2	3
E 05_Topic not present	137	847

5.1.2 Definition of the Building Blocks of the Strategic Fit Matrix

The following table includes the definitions of all the building blocks constituting the strategic fit matrix. They are listed in order of size, from 1 to 14.

D1 financial economic strategies	Productivity strategy: Productivity improvements can occur according to two ways. First, companies reduce costs by lowering direct and indirect expenses. Second, companies by utilizing their financial and physical assets more efficiently *reduce* the working and fixed capital needed to support a given level of business.	Kaplan, R.S. and Norton, D.P. (2004), Strategy Maps: Converting Intangible Assets into Tangible Outcomes, Harvard Business School Publishing Corporation, Boston.
	Growth strategy: Companies can generate profitable revenue growth by deepening relationships with existing customers. This enables them to sell more of their existing products or services or additional products and services. Companies can also generate revenue growth by selling entirely new products or expanding revenues by selling to customers in entirely new segments.	Kaplan, R.S. and Norton, D.P. (2004), strategy maps: Converting intangible assets into tangible outcomes, Harvard Business School publishing corporation, Boston.
D2 generic strategies	Cost leadership: A firm focused on cost leadership becomes a low-cost producer within its industry. The sources of cost advantage are various and depend on the structure of the industry. These may include the pursuit of economies of scale, proprietary technology, preferential access to raw	Porter, M.E. (1985), competitive advantage: Creating and sustaining superior performance, free press, New York.

(continued)

	materials and further factors. If a firm can achieve and sustain overall cost leadership, then it will become an above-average performer in its industry provided it can control prices at or near the industry average.	
	Differentiation: A company can outperform rivals only if it can establish a difference that it can preserve. In order to pursue a differentiation strategy, a company must do the following: • Deliver greater value to customers. • Create comparable value at a lower cost. • Do both. Differentiation arises from both the choice of activities and how they are performed. A company's ability to differentiate itself reflects the contribution of each value-added activity towards the fulfilment of customer needs.	Porter, M.E. (1996), "what is strategy?", Harvard business review, vol. 74 no. 6, pp. 61–78. Porter, M.E. (1985), competitive advantage: Creating and sustaining superior performance, free press, New York.
	Domination (lock-in): The long-term dominant position can be achieved through the specific performance attributes of the company's products and services that describe how the performance of the new product or service dominates the performance of competitors. In order to remain dominant, the company must continually strive to enhance the specific performance dimensions. A key success factor in becoming dominant is to exploit early mover advantages by involving a critical mass of buyers and sellers faster than competitors establishing an industry standard. If the company does not continue to enhance its offering, a rival has the opportunity to offer a superior product or service whose ease and value of use are sufficiently better to overcoming the users' costs switching to the new product. Lock-in arises when companies create high switching costs for their customers. In this situation, one or two companies will tend to be dominant,	Kaplan, R.S. and Norton, D.P. (2004), strategy maps: Converting intangible assets into tangible outcomes, Harvard Business School publishing corporation, Boston.

(continued)

	and they will create large barriers to entry for other exchange providers and high switching costs for its buyers and sellers. Lock-in strategies may not be feasible in all circumstances. Generally, opportunities for lock-in arise only at certain times and in certain segments of industries. But companies should be careful as to when such a strategy might be feasible since the advantages related to a successful lock-in strategy can be enormous.	
D3 value disciplines	Operational excellence: Operational excellence means providing customers with reliable products or services at competitive prices and deliver them with minimal difficulties or inconveniences. Such an effort will be translated in "no frills" solutions and standardized products, as in the following: • Products are competitive. • Distribution is easily managed.	Treacy, M. and Wiersema, F. (1993), "customer intimacy and other value discipline", Harvard business review, vol. 71 no. 1.
	Product leadership: Product leadership means offering customers leading-edge products and services that consistently enhance the customer's use or application of the product, thereby making rivals' goods obsolete. Companies that try to reach product leadership want to fulfil customer needs through excellent performances and superior features.	Treacy, M. and Wiersema, F. (1993), "customer intimacy and other value discipline", Harvard business review, vol. 71 no. 1.
	Customer intimacy: Customer intimacy means segmenting and targeting markets precisely and then tailoring the offer to match exactly the demands of those niches. Companies that excel in customer intimacy combine detailed customer knowledge with operational flexibility so they can respond quickly to almost any need, from customizing a product to fulfilling special requests. As a consequence, these companies engender tremendous customer loyalty. This set of activities requires a personalized solution for each customer.	Treacy, M. and Wiersema, F. (1993), "customer intimacy and other value discipline", Harvard business review, vol. 71 no. 1.
	Ecosystem leadership: In a business ecosystem, companies coevolve	Moore J. F. (1993), "predators and prey: A new ecology of

(continued)

	capabilities around a new innovation: they work together competitively to support new products, fulfil customer needs and eventually incorporate the next round of innovations. While the centre may shift over time, the role of the ecosystem leader is valued by the rest of the community. Such leadership enables all ecosystem members to invest towards a shared future in which they anticipate profit opportunities together. A company should be viewed not as a member of a single industry but as part of a business ecosystem that crosses a variety of industries.	competition", Harvard business review, vol. 11, no. 1.
D4 value proposition	Price: The price can be defined as the total cost incurred by customers to acquire the company's product or service. Efficient operations enable companies to earn a margin on sales sufficient to recovering costs and earning a return on invested capital while still offering customers attractive prices. Companies that strive to become low-cost producers and suppliers would want to be profitable at the lowest price point in the industry. A culture of continuous cost reduction would be relevant for low-cost competitors, especially with non-differentiated products.	Kaplan, R.S. and Norton, D.P. (2004), strategy maps: Converting intangible assets into tangible outcomes, Harvard Business School publishing corporation, Boston.
	Reliability: To compete on reliability, it is necessary to establish a culture of quality and continuous improvement. Customers value rapid, reliable delivery of goods and services. Reliable delivery means delivering on time based on the customer's expectation of the time frame according to which the delivery should occur.	Kaplan, R.S. and Norton, D.P. (2004), strategy maps: Converting intangible assets into tangible outcomes, Harvard Business School publishing corporation, Boston.
	Accessibility, availability and rapidity: This value proposition concerns the time at which the product functionalities become available to customers. Offering excellent functionalities but getting to the market several months or years after competitors will not build the basis for high margins. Companies that excel in delivering this value proposition bring their products and services to the	Kaplan, R.S. and Norton, D.P. (2004), strategy maps: Converting intangible assets into tangible outcomes, Harvard Business School publishing corporation, Boston.

(continued)

	market faster than competitors. Being first-to-market with new products and services is an important source of revenue and margin growth for product leaders.	
	Quality: A company that competes on quality should strive to give to its customers a zero defects experience with its products or services. This aim is straightforward. The ideal customer experience is a product that meets customer specifications and is suited for its immediate use by the customer.	Kaplan, R.S. and Norton, D.P. (2004), strategy maps: Converting intangible assets into tangible outcomes, Harvard Business School publishing corporation, Boston.
	Performance: This component represents the specific performance attributes of the company's products and services. A company's value proposition is focused on performance when the performance of the product or service dominates the competitive offerings. The company must describe and communicate the most important functionalities of the newly introduced products and services. Moreover, it must continuously strive to enhance it.	Kaplan, R.S. and Norton, D.P. (2004), strategy maps: Converting intangible assets into tangible outcomes, Harvard Business School publishing corporation, Boston.
	Newness: Newness is strictly connected with innovation. Successful innovation drives customer acquisition and growth, margin enhancement and customer loyalty. Without innovation, a company's value proposition can eventually be imitated, leading to competition solely on price for its now commoditized products and services. Companies create considerable competitive advantages when they have the capability to bring innovative products, best suited to target customer needs and expectations, on the market fast and efficiently. Companies that strive to maintain product leadership should establish a culture of creativity and product innovation.	Kaplan, R.S. and Norton, D.P. (2004), strategy maps: Converting intangible assets into tangible outcomes, Harvard Business School publishing corporation, Boston.
	Involvement and sharing: Involvement and sharing means to build strong community partnerships, therefore supporting the community involvement identified in the customer perspective. Brand image is driven by these goals. Beyond their inherent value, integrity	Kaplan, R.S. and Norton, D.P. (2004), strategy maps: Converting intangible assets into tangible outcomes, Harvard Business School publishing corporation, Boston.

(continued)

	and fairness fulfil core customer expectations.	
	Product range and customization: This dimension of the value proposition is related to the range of products, merchandise and services offered to customers. Some companies want to be like supermarkets, offering customers a full range of products and services. Other companies, especially those who wish to offer their products to their customers at the lowest price in their industries, recognize that they may not be able to offer every product or service and still execute in a superb way.	Kaplan, R.S. and Norton, D.P. (2004), strategy maps: Converting intangible assets into tangible outcomes, Harvard Business School publishing corporation, Boston.
	Complementary goods, services and experiences: A value proposition based on complementary goods, services and experiences provides the supply of valuable products and services that in order to be used, they require the company's proprietary product or standard. The drivers for push complementary are the services and benefits offered by the company. An example could be the revenues earned from the complements from the use of the product and the company's main technology. A factor of retention is the size of the investment that customers have made in the product or technology owned by the company. The investment could include the training and skills of compliant employees in proprietary technology, as well as specific complementary investments in software, equipment and research and development.	Kaplan, R.S. and Norton, D.P. (2004), strategy maps: Converting intangible assets into tangible outcomes, Harvard Business School publishing corporation, Boston.
	Life (work) style: This dimension concerns the employees and their role in the organization. The culture must encourage generating new ideas and solutions for process improvement and sharing these ideas with other workgroups in the organization. Employees must enthusiastically search for new process improving customer service ideas, whether these ideas arise inside or outside the organization.	Kaplan, R.S. and Norton, D.P. (2004), strategy maps: Converting intangible assets into tangible outcomes, Harvard Business School publishing corporation, Boston.

(continued)

D5 resources	Tangible assets: We can distinguish between tangible and intangible assets. All assets and structures, whether tangible products or intangible, are the results of human action and depend ultimately on people for their continued existence. Tangible means include warehouses, vehicles, production facilities, etc.; intellectual capital is an intangible asset.	Sveiby, K.E. (1997), "the intangible assets monitor", journal of Human Resource Costing & Accounting, vol. 2 no. 1.
	Structural capital: The essence of structural capital is represented by knowledge embedded within the routines of an organization. Its scope is internal to the firm but external to human capital.	Bontis N. (1998), "intellectual capital: An exploratory study that develops measures and models", management decision, vol. 36 issue: 2.
	Human capital: Human capital has been defined as the combination of the following four factors: • Genetic inheritance • Education • Experience • Attitudes regarding life and business The essence of human capital is the sheer intelligence of the organizational member. The scope of human capital is limited to the knowledge node (i.e. internal to the mind of the employee).	Bontis N. (1998), "intellectual capital: An exploratory study that develops measures and models", management decision, vol. 36 issue: 2.
	Relational capital: The essence of relational capital is represented by knowledge embedded in relationships external to the firm. Its scope is external to the firm and internal to human capital. The main topics related to relational capital are the following: • Knowledge of marketing channels • Customer relationships	Bontis N. (1998), "intellectual capital: An exploratory study that develops measures and models", management decision, vol. 36 issue: 2.
	Environmental capital: Environmental capital is part of intellectual capital. Firms that manage environmental capital are more flexible and adaptable to new realities than their competitors, because the creation, transfer and application of knowledge allow them to offer on the market products and services that are more environmentally friendly and therefore have a greater value-added component. The "gap"	Claver-Cortés, E. López-Gamero, M. D. Molina-Azorín, J. F. Zaragoza-Sáez, P. D. C. (2007) "intellectual and environmental capital", journal of intellectual capital, vol. 8 issue: 1.

(continued)

	between the market value and book value of the assets is also known as environmental capital and will increase the value of the firm thanks to the intangible assets created by knowledge management related to the incorporation of the environmental variable.	
D6 suppliers	Market: A company should turn to the market for items where the company plays a dominant market role and suppliers' strength is rated medium or low. If the supply risk is slight, the company has a better chance of achieving a positive profit contribution through favourable pricing and contract agreements. Even so, it has to pay attention not to exploit the advantage so aggressively that it jeopardizes long-term supplier relationships or provokes counterreactions by insisting on rock-bottom prices in times of market discontinuity.	Kraljic P. (1983), "purchasing must become supply management", Harvard business review.
	Supply management (integrated companies): Whenever a manufacturer must procure a volume of critical items competitively under complex conditions, supply management is relevant. The greater the uncertainty of supplier relationships, technological developments and/or physical availability of those items, the more important supply management becomes. A company's need for a supply strategy depends on the following two factors: • The strategic importance of purchasing in terms of the value added by a product line, the percentage of raw materials in total costs and their impact on profitability and so on • The complexity of the supply market gauged by supply scarcity, pace of technology and/or material substitution, barriers to entry, logistic costs or complexities and monopoly or oligopoly conditions	Kraljic P. (1983), "purchasing must become supply management", Harvard business review.
	Technological process interpreters: The technological process interpreters are the producers and suppliers of components and materials that	Verganti, R. (2009), design-driven innovation: Changing the rules of competition by radically innovating

(continued)

	cooperate with the company, succeeding in realizing technology epiphanies. Technology suppliers represent a category of cultural interpreters within the technical world. New technologies, especially the real meanings quiescent within those same technologies, often trigger design-driven innovations.	what things mean, Harvard business press., Boston.
	Technological product interpreters: These interpreters are those hired by the company during the development of pioneering projects. The technical innovations and the new products and services actually change the world of things and, implicitly or explicitly, propose new meanings. In many sectors, innovators use products in special contexts outside the normal market flow, where they have greater freedom to explore new solutions.	Verganti, R. (2009), design-driven innovation: Changing the rules of competition by radically innovating what things mean, Harvard business press., Boston.
	Cultural interpreters: These interpreters explore the culture and meaning behind an explicit component of their core mission. They can be artists, cultural organizations, sociologists, anthropologists, marketers and the media. • Cultural organizations, such as associations, foundations and museums, can promote novel visions and interpretations of meanings, which often contrast with dominant and institutionalized culture. And to this end, they perform research and experiments: • Consultants and researchers in sociology and anthropology of consumption, in marketing, in branding and in communication. With their expertise in observing and analysing society, culture and markets, such interpreters can rapidly provide a picture of how people attribute meanings to things and of emerging trends.	Verganti, R. (2009), design-driven innovation: Changing the rules of competition by radically innovating what things mean, Harvard business press., Boston.
	Technological connection interpreters: These interpreters are research and educational institutions, such as schools, universities and public research organizations that investigate	Verganti, R. (2009), design-driven innovation: Changing the rules of competition by radically innovating what things mean, Harvard business press., Boston.

(continued)

	the evolution of technologies and their impact on culture and society. People conduct such investigations not only in technical schools but also in schools of design, which explicitly address the interactions between technology, languages and meanings. Firms are eager to cooperate with schools of design by challenging professors and students with innovative projects, promoting workshops, hiring young scholars for internships and sharing labs and resources.	
	Ecosystemic actors: They are companies in other industries that address the same life context that a firm is targeting. Every company, in any context, is surrounded by these kinds of interpreters. They can be identified searching which other companies in other sectors are targeting the same people in the same life context and what types of other products or services these people use or could use. All these interpreters have some knowledge of the meanings and languages we are studying. And they would probably be eager to share it and understand our interpretations, since they face the same problems and have the same interests.	Verganti, R. (2009), design-driven innovation: Changing the rules of competition by radically innovating what things mean, Harvard business press., Boston.
D7 processes	Operational management processes: Processes aimed at producing and delivering goods and services to customers. While operational excellence alone is not the basis of a sustainable strategy, managing operations remains a priority for organizations. Operations management processes can encompass up to four important processes as follows: • Developing and sustaining relationships with suppliers • Producing products and services • Distributing and delivering products • Providing services to customers and managing risks	Kaplan, R.S. and Norton, D.P. (2004), strategy maps: Converting intangible assets into tangible outcomes, Harvard Business School publishing corporation, Boston.
	Product development processes: Processes aimed at sustaining a competitive advantage through their continuous innovation to create new	Kaplan, R.S. and Norton, D.P. (2004), strategy maps: Converting intangible assets into tangible outcomes,

(continued)

	products, services and processes. Companies create considerable competitive advantage when they have the capability to bring innovative products, best suited to target customer needs and expectation, on the market fast and efficiently. Product innovation is a prerequisite for participation in technologically based industries, such as the pharmaceutical, the semiconductor and the telecommunication industries. Managing innovation includes the following four important processes: • Identifying opportunities for new products and services • Managing the research and development portfolio • Designing and developing new products and services • Bringing new products and services to market	Harvard Business School publishing corporation, Boston.
	Market development process: Processes focused on transactions, promoting and selling the enterprise's products. Customer management processes must help the company acquire, sustain and grow long-term, profitable relationships with target customers. Customer management consists of the following four generic processes: • Selecting customers: Identifying an attractive customer segment, crafting the value proposition to appeal to these segments and creating a brand image that attracts customers in these segments to the company's products and services • Acquiring customers: Communicating the message to the market, securing prospects and converting prospects to customers • Retaining customers: Ensuring quality, correcting problems and transforming customers into highly satisfied raving fans • Growing relationship with customers: Getting to know customers, building relationships with them and increasing the company's share of target customer purchase activities	Kaplan, R.S. and Norton, D.P. (2004), strategy maps: Converting intangible assets into tangible outcomes, Harvard Business School publishing corporation, Boston.

(continued)

	Social development process: Processes intended to earn the right to operate in the communities and countries in which the company produces and sells. In order to avoid shutdowns or expensive litigations, companies must comply with all these regulations related to business practices. Many companies, however, seek to go beyond complying with the minimal standards established by regulations. They wish to perform better than the regulatory constraints, so that they develop a reputation as an employer of choice in every community in which they operate. Companies manage and report their regulatory and social performance along the following dimensions: • Environmental • Safety and health • Employment practices • Community investment	Kaplan, R.S. and Norton, D.P. (2004), strategy maps: Converting intangible assets into tangible outcomes, Harvard Business School publishing corporation, Boston.
D8 distribution	Extensive: Extensive distribution requires the use of multiple channels in order to increase market penetration and raise barriers to entry. However, intra-brand competition and intra-channel conflicts may become major issues, leading to a reduction in the firm's direct and indirect channels. The basic trade-off in the choice between fewer and more channels appears to be market coverage vs. intermediary investments and value added. Adding more channels may increase market coverage but may reduce the intermediary incentive to invest and add value to the core product.	Frazier, G. L. (1999), "organizing and managing channels of distribution", journal of the academy of marketing science, vol.27, no. 2.
	Selective: When a highly selective distribution approach is adopted, the intent is normally to provide territorial protection to intermediaries to promote their investment in the brand. By reducing intra-brand competition, inter-brand competition may be promoted. Furthermore, monitoring and enforcement efforts appear critical for firms using an exclusive or selective distribution approach, as unauthorized sales outside of assigned territories can cause considerable disruption.	Frazier, G. L. (1999), "organizing and managing channels of distribution", journal of the academy of marketing science, vol.27, no. 2.

(continued)

	Exclusive: Exclusive distribution occurs when the exclusive right to sell the product in a certain area is given to a distributor. The use of exclusive representatives or agents guarantees producers sales channels with low fixed costs and risks. Furthermore, control over these exclusive representatives is very high.	Frazier, G. L. (1999), "organizing and managing channels of distribution", journal of the academy of marketing science, vol.27, no. 2.
	Direct: Direct distribution occurs when the company distributes its products through directly controlled sales channels. Direct channels should be implemented when specialized investments are required to consummate transactions and performance ambiguity is high (Frazier, 1999, p.232). Such factors such as the financial resources of the company and its core competencies, the importance of the product, the market and the customer characteristics (e.g. sizes, needs and preferences) are likely to affect reliance on direct or indirect channels.	Frazier, G. L. (1999), "organizing and managing channels of distribution", journal of the academy of marketing science, vol.27, no. 2.
D9 product	Commodity: True commodities are materials extracted from the natural world: animal, mineral and vegetable. After slaughtering, mining or harvesting the commodity, companies generally process or refine it to yield certain characteristics and then store it in bulk before distributing it to market. By definition, commodities are fungible: they are what they are and therefore interchangeable. Because commodities cannot be differentiated, commodity traders sell them largely into nameless markets where a company purchases them for a price according to supply and demand.	Pine, B.J. and Gilmore, J.H. (1999), the experience economy, Harvard business review press, Boston.
	Goods: Goods are tangible items sold to largely anonymous customers who buy them off the shelf, off the lot, out of the catalogue or on the Web. Because manufacturing processes convert the raw materials in a variety of goods, leeway serves to set prices based on the costs of production as well as on product differentiation.	Pine, B.J. and Gilmore, J.H. (1999), the experience economy, Harvard business review press, Boston.

(continued)

	Services: Services are intangible activities customized to the individual requests of known clients. Service providers use goods to perform operations on a particular client (such as haircuts or eye exams) or on his/her property or possessions (such as lawn care or computer repair). Clients generally value more the benefits of services than the goods required to provide them. Services accomplish specific tasks clients want to be done but do not want to do them themselves; goods merely supply the means.	Pine, B.J. and Gilmore, J.H. (1999), the experience economy, Harvard business review press, Boston.
	Experiences: Experience offerings occur whenever a company intentionally uses services as the stage and goods as props to engage an individual. Experiences have necessarily emerged to create new value. Whereas commodities are fungible, goods are tangible, and services are intangible, while experiences are memorable.	Pine, B.J. and Gilmore, J.H. (1999), the experience economy, Harvard business review press, Boston.
	Transformations: When you customize an experience, you automatically turn it into a transformation, which companies create on top of experiences just as they create experiences on top of services and so forth. A transformation is what the out-of-shape individual, the emotionally troubled person, the young managers, the hospital patient and the struggling company all really desire. By staging a series of experiences, companies are better able to produce a lasting effect on the buyer than through an isolated event. It is the revisitation of a recurring theme, experienced through distinct and yet unified events, that transforms. As multiple experiences emerge and compete for guests, companies staging these events will begin to realize that any experience can become the basis for a new offering that elicits a transformation.	Pine, B.J. and Gilmore, J.H. (1999), the experience economy, Harvard business review press, Boston.
D10 market	Mass market: A company focused on mass markets does not distinguish between different customer segments. The value propositions, distribution channels and customer relationships all	Osterwalder, A. and Pigneur, Y. (2010), business model generation, John Wiley & Sons, Inc., New Jersey.

(continued)

	focus on one large group of customers with broadly similar needs and problems.	
	Segmented (diversified, niche): Some companies distinguish between market segments with slightly different needs and problems. Companies targeting niche markets cater to specific, specialized customer segments. The value propositions, distribution channels and customer relationships are all tailored to the specific requirements of a niche market. Such business models are often found in supplier-buyer relationships. For example, many auto parts manufacturers depend on major automobile manufacturers.	Osterwalder, A. and Pigneur, Y. (2010), business model generation, John Wiley & Sons, Inc., New Jersey.
	Tribe: The word "tribe" refers to the division of the market on the basis of archaic values such as a local sense of identification, religiosity, syncretism, group narcissism and so forth. This term is used in anthropology, to describe archaic societies, in which the social order was maintained without the existence of a central power.	Cova, B. and Cova, V. (2002) "tribal marketing: The tribalisation of society and its impact on the conduct of marketing", European journal of marketing, vol. 36 no. 5/6, pp. 595–620.
	Long tail: Long tail strategy involves selling less of more. It means focusing on offering a large number of niche products, each of which are seldom sold. However, aggregate sales of niche items can be as profitable as the sales of a small number of bestseller products. In order to be able to approach a large number of niches in a profitable manner, it is necessary to keep inventory costs low and have strong platforms to make niche content readily available to interested buyers.	Osterwalder, A. and Pigneur, Y. (2010), business model generation, John Wiley & Sons, Inc., New Jersey.
D11 clients	Convenience, functionality and usability: They are directed to satisfy what Maslow, 1943, defines as physiological needs. Physiological drives or needs are isolable and are localizable somatically. They are relatively independent of each other, of motivations and of the organism. In many cases, it is possible to demonstrate a localized, underlying somatic base for the driver. This is true less generally than has been thought	Maslow, A. H. (1943), "A theory of human motivation", psychological review, vol. 50, pp. 370–396.

(continued)

	(exceptions are fatigue, sleepiness and maternal responses), but it is still true in the classic instances of hunger, sex and thirst.	
	Security: This dimension includes companies that aim to satisfy the safety and security needs. They arise when the physiological needs are relatively fulfilled, and then a new set of needs emerges, which we can categorize roughly in safety and security needs. The healthy, normal, fortunate adult in our culture is largely satisfied in terms of his/her safety and security needs. The peaceful, smoothly running, "good" society ordinarily makes its members feel safe enough from wild animals, extreme temperatures, criminals, assaults and murders, tyranny, etc.	Maslow, A. H. (1943), "A theory of human motivation", psychological review, vol. 50, pp. 370–396.
	Sense of belonging and self-esteem: Sense of belonging: If both the physiological and the safety and security needs are fairly fulfilled, then the sense of belonging, love and affection needs will emerge. The person will feel, as never before, the absence of friends or a sweetheart, or a wife or children, keenly. He/she will hunger for affectionate relationships with people in general, namely, for a place in the group, and he/she will strive with great intensity to achieve this goal. He/she will want to attain such a place more than anything else in the world and may even forget that once; when he/she was hungry, he/she sneered at love. Self-esteem: All people in our society (with a few pathological exceptions) have a need or desire for a stable, firmly based (usually), high evaluation of themselves, for self-respect, for self-esteem and for other esteems. By firmly based self-esteem, we mean that which is soundly based upon real capacity, achievement and respect from others. These needs may be classified into two subsidiary sets. These include, firstly, the desire for strength, for achievement, for adequacy, for confidence in the face of the world and for independence and freedom. Secondly, we have what we	Maslow, A. H. (1943), "A theory of human motivation", psychological review, vol. 50, pp. 370–396.

(continued)

	may call the desire for reputation or prestige (defining it as respect or esteem from others), recognition, attention, importance or appreciation.	
	Aesthetic and creativity: Creativity and aesthetic both represent the need to go beyond self-affirmation, which is mainly guided by the ego, in self-realization through far-reaching connections with the world in which we live. Creativity represents the courage to open yourself to the world and interact directly with it. True creativity is always accompanied by a keen sense of conversation and exchange with the world in general.	Maslow, A. H. (2012), toward a psychology of being, start publishing LLT, United States
	Knowledge and understanding: It is a basic desire to know, to be aware of reality, to get the facts, to satisfy curiosity or, as Wertheimer claims, "to see rather than to be blind". This need is difficult to satisfy because we know so little about the facts. Curiosity, exploration, desire for facts and desire to know may certainly be observed easily enough.	Maslow, A. H. (1943), "A theory of human motivation", psychological review, vol. 50, pp. 370–396.
	Self-realization need/transcendence: Self-actualization need: It refers to the desire for self-fulfilment, namely, the realization of a person's potential. This tendency might be phrased as the desire to become more and more of what one is and to become everything that one is capable of becoming. Transcendence: Transcendence refers to the highest and most inclusive or holistic level of human consciousness, behaving and relating, as ends rather than means, to oneself, to significant others, to human beings in general, to other species, to nature and to the cosmos.	Maslow, A. H. (1943), "A theory of human motivation", psychological review, vol. 50, pp. 370–396. Maslow, A. H. (1971), the farther reaches of human nature, penguin, New York.
	Useful: The word "useful" is to be understood as used here in a linguistic or metaphorical sense. When we say that water is useful for putting out fire, the word "useful" is used in a non-scientific sense. Water thrown on the fire is the cause of it going out: this is the knowledge that serves for basis to the action, let us say, of firemen. There	Croce, B. (1922), aesthetic as science of expression and general linguistic, Macmillan, London.

(continued)

	is a link, not of nature, but of simple succession, between the useful action of the person who extinguishes the conflagration and this knowledge. The technique of the effects of water is the theoretical activity which precedes; the *action* of him/her who extinguishes the fire is alone useful.	
	Beautiful: Beauty is the visible idea of perfection. It falls under the influence of the design topic. There is a multiplicity of beauties. Things are beautiful according to our ideas about them. So, every beautiful thing has its own kind. He adds to his observations in this sense: "Since the diamond is only perfect among stones, gold among metals and man among living creatures, so there is a distinction in each species, and little is perfect". Beauty is an intermediate arrangement, which contains a part of perfection and a part of the pleasant, and forms a tertium quid, which differs from the other two and deserves a special name.	Croce, B. (1922), aesthetic as science of expression and general linguistic, Macmillan, London.
	Right and good: In order to resolve to perform an action, it is necessary to have judged whether "this action is useful or this action is good". A good action is an action that is wanted. It will always be impossible to distil from the objective study of things a single drop of goodness. We do not desire things because we know them to be good or useful; but we know them to be good and useful, because we desire them. Practical action is preceded by knowledge, but not by practical knowledge, or better by practice: to obtain this, it is first necessary to have practical action.	Croce, B. (1922), aesthetic as science of expression and general linguistic, Macmillan, London.
D12 society	Re(configuring) operations to improve the social impact: For creating this kind of shared value, it is necessary to identify all the societal needs that a company's value chain inevitably affects and is affected by. There are numerous societal issues, such as natural resource and water use, health and safety, working conditions and equal treatment in the workplace.	Porter, M.E. and Kramer, M.R. (2011), creating shared value, how to reinvent capitalism and unleash a wave of innovation and growth, Harvard business review, vol. 8, no. 1–2.

(continued)

	Opportunities to create shared value arise because societal problems can create economic costs in the firm's value chain. Many so-called externalities actually inflict internal costs on the firm, even in the absence of regulation or resource taxation.	
	Redesigning products and markets to meet the social needs: For a company, the starting point for creating this kind of shared value is to identify all the societal needs, benefits and harms that are or could be embodied in the firm's products. The opportunities are not static; they change constantly as technology evolves, economies develop and societal priorities shift. An ongoing exploration of societal needs will lead companies to discover new opportunities for differentiation and repositioning in traditional markets and to recognize the potential of new markets they previously overlooked. Meeting needs in underserved markets often requires redesigned products or different distribution methods. These requirements can trigger fundamental innovations that also have application in traditional markets.	Porter, M.E. and Kramer, M.R. (2011), creating shared value, how to reinvent capitalism and unleash a wave of innovation and growth, Harvard business review, vol. 8, no. 1–2.
	Supporting local ecosystem development: The success of every company is affected by the supporting companies and infrastructure around it. It is the ecosystem. Productivity and innovation are strongly influenced by ecosystem or geographic concentrations of firms, related businesses, suppliers, service providers and logistical infrastructure in a particular field. Ecosystems include not only businesses but institutions such as academic programmes, trade associations and standards organizations. They also draw on the broader public assets in the surrounding community, such as schools and universities, clean water, fair competition laws, quality standards and market transparency. Clusters are prominent in all successful and growing regional economies and play a	Porter, M.E. and Kramer, M.R. (2011), creating shared value, how to reinvent capitalism and unleash a wave of innovation and growth, Harvard business review, vol. 8, no. 1–2.

(continued)

	crucial role in driving productivity, innovation and competitiveness. Firms create shared value by building ecosystems to improve company productivity while addressing gaps or failures in the framework conditions surrounding the ecosystem.	
D13 costs	Costs for the volume produced: The volume-based costs are costs that increase or decrease in relation to the amount of product produced or the quantity of service provided. Good examples of volume-based costs are direct material costs and direct labour costs. These costs increase with each unit of the volume of output.	Blocher, E. J. stout, D. E. and Cokins, G. (2010), cost management a strategic emphasis, McGraw-hill/ Irwin, New York.
	Costs for production capacity: The costs for production capacity are fixed costs that provide capacity for operations. Good examples of production capacity costs are plant building and equipment.	Blocher, E. J. stout, D. E. and Cokins, G. (2010), cost management a strategic emphasis, McGraw-hill/ Irwin, New York.
	Costs for research and development (R&D): Research and development (R&D) costs are expenditures aimed at developing new products and processes or at modifying existing products or processes. Examples of R&D costs include laboratory research aimed at discovering new knowledge; searching for applications of new research findings or further knowledge, conceptual formulation and design of possible product or process alternatives; and testing in search for or the evaluation of product or process alternatives.	Hansen, D.R. Mowen, M.M. and guan, L. (2008), cost management accounting and control, South-Western Cengage learning, Mason.
	Costs for training and development: Training and development costs usually are included within the indirect labour costs. Indirect labour costs also include supervision, quality control, inspection, purchasing and receiving, material handling, janitorial labour, downtime, training and cleaning-up.	Blocher, E. J. stout, D. E. and Cokins, G. (2010), cost management a strategic emphasis, McGraw-hill/ Irwin, New York.
	Costs for marketing and communication: The costs that are necessary to market and distribute a product or service are marketing (selling) costs. They are often referred to as order-getting and order-filling	Hansen, D.R. Mowen, M.M. and guan, L. (2008), cost management accounting and control, South-Western Cengage learning, Mason.

(continued)

	costs. Examples of marketing costs include sales personnel salaries and commissions, advertising, warehousing, shipping and customer service. The first two items are examples of order-getting costs; the last three are order-filling costs.	
	Costs for relationship management: These costs are the costs required for the maintenance of relationships between the company and the actors involved in its micro-environment including suppliers, market intermediaries, the public and of course customers who constitute the most important actors. In relation to customers, the idea is first and foremost to enhance customer loyalty as well as to create a stable, mutually profitable and long-term relationship.	Ravald, A. Grönroos, C. (1996) "the value concept and relationship marketing", European journal of marketing, vol. 30 issue: 2.
D14 pricing	Economy pricing: Economy pricing occurs when a company sets the price, undercutting the prices of its competitors. Every company that succeeds in growing profitably with a low-price strategy must first create a business model that enables it to cut incremental costs below those of its competitors.	Nagle, T.T. and Müller, G. (2018), the strategy and tactics of pricing: A guide to growing more profitably, Routledge, New York and London.
	Penetration pricing: Penetration pricing involves setting a price low enough to attract and hold a large base of customers. Penetration prices are not necessarily cheap, but they are low relative to perceived value in the target segment.	Nagle, T.T. and Müller, G. (2018), the strategy and tactics of pricing: A guide to growing more profitably, Routledge, New York and London.
	Premium price: A company can obtain a premium price from its products or services when they are sold at a higher price than the reference value thanks to the extra benefits that these guarantee. The reference value is the estimate of the monetary value (Nagle and Müller, 2018, pp. 109–110). Premium product marketing experts will review premium prices in a different context to reduce buyer price sensitivity.	Nagle, T.T. and Müller, G. (2018), the strategy and tactics of pricing: A guide to growing more profitably, Routledge, New York and London.
	Loyalty pricing is designed to capture superior margins, even at the expense of high sales volume. By definition, skimming pricing occurs when prices	Nagle, T.T. and Müller, G. (2018), the strategy and tactics of pricing: A guide to growing more profitably, Routledge, New York and London.

(continued)

	are high in relation to what most buyers in a segment can be convinced to pay. Consequently, this strategy optimizes immediate profitability only when the profit from selling products to relatively price-insensitive customers exceeds that from selling to a larger market at a lower price. In some instances, products might reap more profit in the long run by setting initial prices high and reducing them over time even if those high initial prices reduce immediate profitability. Sequential skimming, like the more sustainable variety of skimming, begins with a price that attracts the least price-sensitive buyers first. Once the firm has created its cream skimming effect, however, that market is gone. Consequently, to maintain its sales, the firm must reduce its price enough to sell to the next most lucrative segment. The firm continues this process until it has exhausted all segments with profitable volume potential. In theory, a firm could sequentially skim the market for a durable good or a one-time purchase by lowering its price in hundreds of small steps, thus charging every segment the maximum it would pay for the product.	
	Product bundle pricing: Bundle pricing occurs when a seller segments the market by configuring different offers consisting of different packages of features and services for different segments. In bundle pricing, companies sell a package or a set of goods or services for lower prices than they would charge if they bought all of them separately. To create an effective aggregate offer, you must first determine what features and services the company must include in its packages. Savings for customers and companies derive from the fact that a single price for a set of features and services reduces transaction costs for customers and sellers. Finally, the research shows that people are less sensitive to the cost of value-added features and services when grouped together as a single expense.	Nagle, T.T. and Müller, G. (2018), the strategy and tactics of pricing: A guide to growing more profitably, Routledge, New York and London.